STRATHCLYDE UNIVERSITY LIBRARY

30125 00089445 0

KU-384-825

This book is to be returned on or before
the last date stamped below.

28 JAN 1987
12 MAR 1987
30 APR 1987

LIBREX —

Introduction to Fluid Logic

SERIES IN THERMAL AND FLUIDS ENGINEERING

JAMES P. HARTNETT and THOMAS F. IRVINE, JR., Editors
JACK P. HOLMAN, Senior Consulting Editor

Cebeci and Bradshaw	• Momentum Transfer in Boundary Layers
Chang	• Control of Flow Separation: Energy Conservation, Operational Efficiency, and Safety
Chi	• Heat Pipe Theory and Practice: A Sourcebook
Eckert and Goldstein	• Measurements in Heat Transfer, 2nd edition
Edwards, Denny, and Mills	• Transfer Processes: An Introduction to Diffusion, Convection, and Radiation
Fitch and Surjaatmadja	• Introduction to Fluid Logic
Ginoux	• Two-Phase Flows and Heat Transfer with Application to Nuclear Reactor Design Problems
Hsu and Graham	• Transport Processes in Boiling and Two-Phase Systems, Including Near-Critical Fluids
Hughes	• An Introduction to Viscous Flow
Kreith and Kreider	• Principles of Solar Engineering
Lu	• Introduction to the Mechanics of Viscous Fluids
Moore and Sieverding	• Two-Phase Steam Flow in Turbines and Separators: Theory, Instrumentation, Engineering
Richards	• Measurements of Unsteady Fluid Dynamic Phenomena
Sparrow and Cess	• Radiation Heat Transfer, augmented edition
Tien and Lienhard	• Statistical Thermodynamics, revised printing
Wirz and Smolderen	• Numerical Methods in Fluid Dynamics

PROCEEDINGS

Keairns	• Fluidization Technology
Spalding and Afgan	• Heat Transfer and Turbulent Buoyant Convection: Studies and Applications for Natural Environment, Buildings, Engineering Systems

Introduction to Fluid Logic

E. C. FITCH

Fluid Power Research Center
Oklahoma State University
Stillwater, Oklahoma

J. B. SURJAATMADJA

Mechanical Research and
Development Department
Halliburton Services
Duncan, Oklahoma

HEMISPHERE PUBLISHING CORPORATION
Washington   London

McGRAW-HILL BOOK COMPANY

New York   St. Louis   San Francisco   Auckland   Bogotá
Düsseldorf   Johannesburg   London   Madrid   Mexico
Montreal   New Delhi   Panama   Paris   São Paulo
Singapore   Sydney   Tokyo   Toronto

INTRODUCTION TO FLUID LOGIC

Copyright © 1978 by Hemisphere Publishing Corporation. Preliminary edition copyright © 1975 by E. C. Fitch and J. B. Surjaatmadja. All rights reserved. Printed in the United States of America. No part of this publication may be reproduced, stored in a retrieval system, or transmitted, in any form or by any means, electronic, mechanical, photocopying, recording, or otherwise, without the prior written permission of the publisher.

1 2 3 4 5 6 7 8 9 0   K P K P   7 8 3 2 1 0 9 8

This book was set in Press Roman by Hemisphere Publishing Corporation.
The printer and binder was The Kingsport Press, Inc.

Library of Congress Cataloging in Publication Data

Fitch, Ernest C
   Introduction to fluid logic.

   Includes bibliographies and index.
   1. Fluid logic.   2. Logic design.   I. Surjaatmadja, J. B., joint author.   II. Title.
TD853.F55   629.8′04′2   77-24504
ISBN 0-07-021126-4

Contents

Preface ix

Chapter 1 INTRODUCTION 1

    1-1 What Is Fluid Logic? 1
    1-2 Evolution of Fluid Logic 2
    1-3 Why Fluid Logic? 3
    1-4 Industrial Applications 5
    1-5 Gist of Fluid Logic Design 7
        Definitions 8
        References 9

Chapter 2 BOOLEAN ALGEBRA 10

    2-1 The Algebra 10
    2-2 Boolean Theorems 11
    2-3 Truth Table Proof 12
    2-4 The Venn Diagram 13
    2-5 Canonical Forms 20

2-6 Uncomplementation Algorithm   23
2-7 Karnaugh Maps   24
2-8 Computer-aided Simplification   33
     Problems   35
     Definitions   38
     References   39

Chapter 3 SYSTEM INPUTS AND OUTPUTS   40

3-1 Interconnection Philosophy   40
3-2 Input Sensors   41
3-3 Input Interface and Amplification   48
3-4 Timing Devices   51
3-5 Output Circuits   54
3-6 The Power and Input Systems   55
3-7 Application Notes   59
     Problems   63
     Definitions   67
     References   67

Chapter 4 LOGIC ELEMENTS AND CIRCUITS   69

4-1 Logic Network Implementation   69
4-2 Types of Logic Elements   70
4-3 Implementation Considerations   95
4-4 Equation Implementation Circuits   97
4-5 Fluid Circuits   101
     Problems   109
     Definitions   112
     References   112

Chapter 5 LOGIC SYSTEM DESCRIPTION   114

5-1 Fundamental Consideration   114
5-2 Truth Table   115
5-3 Timing or Bar Chart   116
5-4 The Operations Table   119
5-5 Primitive Flow Table   120
5-6 The State Matrix   124
5-7 The Synthesis Table   126
5-8 The Logic Specification Chart   127
5-9 Construction Mechanics   129
     Problems   133
     Definitions   135
     References   136

Chapter 6 COMBINATIONAL LOGIC DESIGN 137

- 6-1 Design Methodology 137
- 6-2 Conventional Solution 138
- 6-3 Utilization of "Don't Cares" 140
- 6-4 Reduction by Complementation 143
- 6-5 Minimization by Factoring 145
- 6-6 Multiterminal Networks 148
- 6-7 Utilization of Logically Complete Elements 153
- 6-8 Three-Level NOR Logic 159
- 6-9 Combinational Network Hazards 166
- 6-10 Computer-aided Synthesis 172
- Problems 180
- Definitions 183
- References 183

Chapter 7 SEQUENTIAL LOGIC DESIGN—CLASSICAL SYNTHESIS 186

- 7-1 Logic Description 186
- 7-2 State Equivalency 188
- 7-3 State Substitution 191
- 7-4 Reduced Specification Chart 194
- 7-5 Minimal Row Specification Chart 195
- 7-6 Operational Flow Chart 199
- 7-7 Excitation Charts and Maps 207
- 7-8 Output Maps 215
- 7-9 Sequential Network Hazards 217
- 7-10 Computer-aided Synthesis 226
- Problems 227
- Definitions 231
- References 231

Chapter 8 SEQUENTIAL LOGIC DESIGN—NONCLASSICAL 234

- 8-1 Quest for Optimal Synthesis 234
- 8-2 The Change Signal Method 235
- 8-3 The Total Signal Method 245
- 8-4 The State Matrix Method 250
- 8-5 Transition Table Method 253
- 8-6 The State Diagram Method 259
- 8-7 Computer-aided Synthesis 267
- Problems 269
- Definition 273
- References 274

Chapter 9   NETWORK ANALYSIS AND REVISION   275

    9-1   Esoteric Reality of Analysis   275
    9-2   Analysis of Combinational Logic Networks   276
    9-3   Combinational Logic Network Revision   279
    9-4   Sequential Network Analysis   282
    9-5   Sequential Network Revision   293
    9-6   Tabular Method for Network Analysis   298
    9-7   Troubleshooting   302
    9-8   Computer-aided Analysis   303
          Problems   305
          Definitions   307
          References   308

Appendix   A TYPICAL INDUSTRIAL SYNTHESIS PROBLEM   309

          Definitions   317
          Index   321

Preface

This book is an outgrowth of some twelve years' experience with a course in fluid logic at Oklahoma State University. The course is essentially an introduction to an area of engineering science that includes many subjects important for later professional study and applications in industry. Since an introductory textbook must lay an adequate foundation for advanced study and application, the book assumes no specific background in fluid logic or switching algebra.

The emphasis of the writing has been directed toward achieving two objectives: (1) to develop understanding of the basic concepts of fluid logic and (2) to develop proficiency with the synthesis and analysis techniques to satisfy many of the applications in industry. To accomplish these objectives, the book is carefully packaged to serve the goals of various readers equally well. Since the book is oriented toward self-study, the user may enter it at any chapter where previous training permits and proceed as far as requirements or interests dictate. This prevailing feature has resulted in somewhat greater detail in text material, procedures, and illustrative examples. Experience in the classroom has indicated that such features will decrease the teacher's work load by placing more responsibility on the student in the learning process, thus allowing more time to concentrate on in-depth study and areas where students need help.

To make the text meaningful and valuable to those interested in fluid logic, the authors have attempted to unify the various concepts, techniques, and interacting disciplines to produce an orderly exposition on the subject. Abstract notions are kept to a minimum, while stress is placed on connecting ideas and methodology in keeping with physical engineering reality. Thus, the book can serve not only as a teaching medium for engineering and technology students but also as a self-study aid for engineers in practice.

A special effort was made to present applicable computer-aided design techniques. Unique to this book is the identification and illustration of practical computer programs which are available to simplify and eliminate hazards in network equations, to synthesize combinational and sequential-type logic systems, and to analyze logic networks for errors.

The introductory chapter discusses the state of the art of fluid logic and emphasizes areas of potential application. Chapter 2 provides the reader with the bare essentials of Boolean algebra and advances a practical method for simplifying Boolean equations manually and with a computer. Chapter 3 gives a detailed presentation of peripheral equipment, including selection criteria. This chapter also provides the basis for preparing the machine logic specification. Chapter 4 serves as the library of fluid logic elements and represents each type of element in cross-sectional, algebraic, truth table, and symbolic (A.N.S.I. hydraulic, fluidic, and fluid logic) forms. In addition, the two A.N.S.I.-approved diagramming methods (attached and detached) for fluid logic networks are systematically presented.

Against this background, Chapter 5 reviews a rather exhaustive collection of methods which have been found suitable for describing logic specifications and proceeds to introduce a universal format for recording the required logic of all types of logic systems—called a Logic Specification Chart (LSC). Chapter 6 addresses the problem of synthesizing and implementing combinational logic circuits. Hardware minimization techniques are discussed in detail, including those involving NOR elements. Chapter 7 presents the "classical" synthesis technique as conceived originally by Huffman along with various modifications the authors felt were needed to maintain its position as a "referee" method in the future. Chapter 8 deals with the most notable of the nonclassical synthesis techniques, such as tabular and matrix techniques, in order to give the reader an in-depth understanding of the mechanics of fluid logic. The book culminates with a presentation on fluid logic network analysis, including techniques for correcting erroneous logic systems without resorting to resynthesis.

A special feature of the book is a running list of logic definitions and terms in addition to a comprehensive list of applicable references at the end of each chapter. To improve the suitability of the book for classroom use, an extensive set of problems is given at the end of each chapter to help the student gain mastery of the subject. Finally, a typical industrial-oriented logic problem is synthesized in the Appendix, using one of the effective techniques covered in the book.

First and foremost, the authors would like to acknowledge the many contributions made by the students each year in their formal course on fluid logic. Second, the advancements in the field of fluid logic reflected by this book are a tribute to

PREFACE

the persistent efforts of a host of graduate students at Oklahoma State University who elected to conduct their research in this field—notable among these are J. A. Caywood, D. M. DeMoss, J. H. Cole, G. E. Maroney, and R. L. Woods.

The authors are particularly grateful to their associates at the Fluid Power Research Center for their individual help and encouragement. Inputs received from industrial fluid logic practitioners such as Rick Olszewski of Parker-Hannifin and Glenn E. Wall of Air Hydraulics Company were especially valuable throughout the development of the book. The publications staff at the Fluid Power Research Center, particularly Lyn Engelhardt, had a major part in preparing the manuscript. Of special note is the talent and craftsmanship of draftsmen/photographers Lynn Alger and Joel Moore, abundantly evident in the many complex illustrations throughout the book. Last, but not least, the authors wish to express their gratitude to their respective families for the years of encouragement and sacrifice they made—it is their book too and it is dedicated to them.

E. C. Fitch
J. B. Surjaatmadja

Introduction to Fluid Logic

Chapter 1

Introduction

1-1 WHAT IS FLUID LOGIC?

Fluid logic is the study of digital-type fluid control elements and the interconnection of such elements to satisfy a given logic specification. Fluid logic circuits utilize a particular class of control elements which possess discrete states. The primary purpose of the fluid logic circuit incorporating these elements is to accomplish a control function rather than a power transmission service. All machine systems must contain some type of digital control in order to provide the necessary on-off or sequential control of machine operations.

Webster defines logic as the "interrelation or sequence of facts or events when seen as inevitable or predictable" and also as "the fundamental principles and the connection of circuit elements for arithmetical computation in a computer." Actually, logic is the science of argument and the transformation of words and thoughts into symbols representing a language which eliminates ambiguities and confusion. The symbolism utilized in fluid logic is derived from mathematical logic and makes use of logic functions which provide an orderly means of representing a logic specification. It is truly a science and one which should be of concern not only to today's engineer but to tomorrow's engineer and manager.

In fluid logic, a fluid (either a liquid, a gas, or both) is used to transmit signals and perform the logic function. Although hybrid systems (those involving other signal media such as mechanical and electrical) are common, a pure fluid system would exhibit fluid inputs, outputs, and circuit elements. Fluid logic elements include both fluidic (nonmoving parts hardware) and those having moving parts such as diaphragms, membranes, balls, and spools. When the logic specification is simple and involves low power levels, the use of fluidic elements has proved rewarding. However, in complex logic networks, the use of air logic (miniaturized, modularized moving-parts-type pneumatic elements) is almost mandatory. The introduction of new network elements and the advancement of various sensors and transducers for providing fluid input signals have contributed immensely to the viability of modern fluid logic.

1-2 EVOLUTION OF FLUID LOGIC

Fluid logic owes its existence to formal logic, which was founded by Aristotle (approximately 400 B.C.) when he formalized his system of syllogisms. There are many reports throughout history of attempts by philosophers to find a manageable symbolism for the formalization of logic. Traditionally the science of the philosopher, logic has only been of interest to the mathematician in the past century or so. In 1854, an English mathematician, George Boole, presented the first practical system of logic in algebraic form. Boole's system established a new mathematics called Boolean algebra in which the problems in logic could be represented and solved in a manner similar to conventional algebra. Since the advancement of algebraic logic by Boole, many mathematicians have made major contributions and have formed what is recognized and employed today as modern Boolean algebra.

Algebraic logic was conceived for the purpose of implementing the solution of logic problems and was not invented with any technical application in mind. For eighty years, symbolic logic was generally regarded as an interesting but useless concept with no practical significance. In 1938, while still a graduate research assistant at M.I.T., Claude E. Shannon recognized that the logical structure of an electrical switching circuit was comparable to the structure of symbolic logic. This structure was that Boolean algebra is a two-valued or binary algebra wherein every term has just two exemplary values with systematic rules for the use of three fundamental connectives, AND, OR, and NOT. Shannon demonstrated the application of classical Boolean algebra by providing an orderly algebraic procedure for the treatment of relay contact networks and thus established himself as the progenitor of modern switching theory.

The theory that resulted from Shannon's work provided the means by which combinational-type logic circuits could be mathematically designed. Combinational logic circuits are characterized by output values which are dependent solely upon a unique set of input values. Through the advancement of Shannon's switching theory, the synthesis of electrical combinational circuits has become a mature field in which all classes of these circuits, including series-parallel, multi-terminal, and nonseries-parallel networks, can be designed in a routine manner.

It was not until 1954 that a mathematical approach was advanced to synthesize sequential circuitry. Such circuits are characterized by outputs which depend not only on the immediate values of the input variables but also on the input history. In 1954, D. A. Huffman and E. A. Moore concurrently and independently developed a method by which the past input conditions of a system could be recorded and become an integral part of the circuit solution. The Huffman-Moore model for sequential circuit synthesis as it is now known is widely used for the design of electrical sequential circuitry and serves as the basis of the reference (classical) procedure for fluid logic sequential circuits today.

H. R. Ronan, in 1959, was credited with the first attempt to achieve the transition from intuitive to logical design of fluid logic circuits. His disclosure demonstrated the use of Boolean algebra for the representation of fluid logic elements and introduced a technique for the design of "hydraulic switching circuits." The significance of Ronan's work today is of little value, but the studies which were precipitated by his presentation will continue to earn him credit for his original contribution.

In general, the formal methods currently available for synthesizing fluid logic networks have stemmed from the ongoing fluid logic program at Oklahoma State University which was initiated in 1956. Over fifty publications document the various concepts and methods generated by this program, and some of the pivotal work is reported in the referenced material at the end of this chapter by Burchett, Fitch, Cole, Maroney, Woods, and Surjaatmadja. Through the contributions of many different investigators in fluid logic at Oklahoma State University, a most sophisticated methodology for the synthesis of such systems has evolved. This book attempts to bring the various facets of fluid logic into perspective in order to provide a suitable foundation for future efforts.

1-3 WHY FLUID LOGIC?

Whenever the combination of high output power and low response time is a consideration, fluid logic is a prime candidate for the logic control network. As illustrated in Figure 1-1, fluid logic excels when circuit response does not exceed 1/100 of the reaction time of a human operator (0.5 seconds). Such a restraint is unimportant for most machine control applications. Only when computational service is involved does the response time limitation of fluid logic suffer.

Fluid logic systems are capable of controlling many tasks. Perhaps the most attractive applications are on machines which utilize fluid power or where high levels of power are required for control purposes. In the case of a fluid power system, a fluid logic network can be connected directly to the power system in order to exercise complete or partial automatic control of the machine's sequence of operation without any conversion of energy media. Combining logic systems with servo-mechanisms and the machine's integral power units can create a control capability which could not only provide regulation for the rate at which a task is performed but could also give the machine the ability to make decisions. The complexity of the decision-making ability of the machine is limited only by the

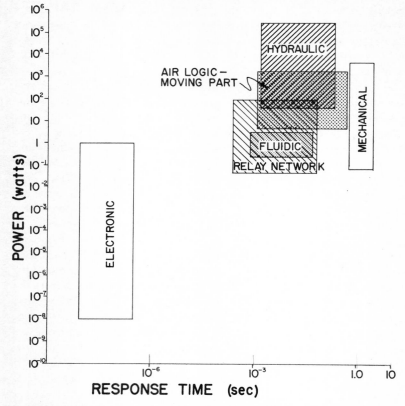

Figure 1-1 Technical comparisons of various types of logic elements.

degree of sophistication of the digital system itself. These control systems can partially or completely replace the need for a human operator.

Spurred on by the advancement and development of miniaturized logic elements, fluid logic is rapidly gaining acceptance and respect throughout industry as a means of controlling automatic operations. Periodicals continue to report on a wide variety of applications for fluid logic. The replacement of electrical controls on assembly-line machines with recently developed air logic systems is commonplace. There is a growing realization that fluid systems are capable of satisfying all the logic functions that conventional electrical relay systems performed in the past—and often at less cost and with greater reliability.

Fluid logic control systems are well suited for operating under adverse conditions. These systems can operate at extremely high or low temperature, are immune to nuclear radiation, and are capable of withstanding extreme vibrations. They offer an increase in reliability over equivalent electrical systems in adverse working conditions and are free from fire and electrical shock hazards. Due to their modularized construction and overall simplicity, fluid logic systems enjoy the reputation of unsurpassed maintainability, reliability, and service life.

INTRODUCTION

Troubleshooting is simple and direct, with no high energy sources to cause a hazardous operation. Of all the control systems which have been devised, no other is more oriented toward heavy industry than fluid logic.

1-4 INDUSTRIAL APPLICATIONS

Fluid logic has been applied in almost every modern industry where the control of dynamic systems must be accomplished effectively, safely, and economically. The utilization of fluid logic in industry will continue at a rapid pace as long as proper logic devices are available and appropriate methodology exists to formulate their interconnection. The application of fluid logic has been particularly acute in those industries where electrical controls are not compatible with the machine or its environment and where reliability on high power level machines is critical.

In the packaging industry, pneumatic logic offers a hygienically clean means for controlling the various operations involved. For example, in the bottling area, the positioning, filling, capping, marking, inspecting, packing, and stacking operations can be accomplished by a single network of air logic circuits. The packaging industry is characterized by special-purpose apparatuses where the existence of mass-produced machines is a rarity. Hence, the need for custom-designed and -fabricated controllers is of paramount importance. This necessity can be appreciated by becoming aware of the broad spectrum of the types of machines which are required to package various products. The following list reflects the kinds of packaging and processing equipment in which fluid logic plays a part:

> dispensers, folders, gluers, tyers, weighers, packers, banders, bundlers, labelers, sealers, wrappers, punchers, shakers, coaters, feeders, perforators, imprinters, gusseters, conveyers, elevators, formers, winders, dispatchers, coders, washers, spreaders, laminators, line dividers and convergers, stretchers, and shrink tunnels.

The aircraft industry has applied fluid logic in both flight and jet engine control areas. Various fluid devices have been advanced to provide attitude control, roll-rate control, and pitch-axis damper control. In jet engine controllers, a means is provided to obtain the maximum possible performance from the engine without causing low-speed unstable surges, compressor stall, or excessive temperatures at high speeds. The controller actually regulates the fuel flow as a direct function of engine temperature, throttle position, aircraft speed, and ambient temperature. Many new applications are continuing to be revealed and developed, including pneumatic anti-skid controllers, wheel stabilizers, and seat ejectors.

The biomedical engineering field contains many examples of life-saving fluid logic devices. The controls for the artificial heart pump and the heart massager represent important medical applications. Many reasons have been given for avoiding electrical apparatuses around patients—the accidental deaths resulting from such equipment are attributed to incompatible electronic devices, high leakage current, poor circuit design, accidental grounding by patient, errors by hospital staff, and poor wiring in hospitals. Fluid logic circuits have been applied to the following medical procedures and treatments:

1. Administration of drugs—orally and intravenously.
2. Exercise and manipulation of various members of the body.
3. Cutting, extraction, repair, and suturing of various tissues or bones.
4. Traction, splinting, pressing, or binding the body.
5. Local or general control of breathing gases, temperature, humidity, and pressure.

Medical science is also using or considering biochemistry procedures to diagnose patient illnesses. In this regard, the development of a pneumatic logic diluter has provided an accurate means of sampling and diluting various fluids. Other fluid logic applications in this field include an automatic catheter flushing system and respirator (both passive and active).

The manufacturing industry has innumerable applications for fluid logic control systems. Basic machine tools which have utilized fluid logic controls include the following:

> automatic lathes, spindle drills, transfer presses, forming machines, tracers, lapping machines, tube benders, indexers, gear hobbers, jig borers, duplex milling machines, key seaters, gage surveillance, broachers, screw machines, wheel dressers, alligator shears, and punching machines.

On machine assembly lines, pneumatic logic networks have been designed to dispatch parts, index and position elements, control the sequence of machine operations, inspect and reject defective parts, coordinate and store parts, and requisition and transfer parts for final assembly. The manufacture of bars, sheets, and plates as well as extrusions, castings, forgings, and moldings has provided endless application areas for fluid logic controllers. In many cases, the environment itself has prohibited the reliable use of any other form of machine control.

In the earth moving and agricultural industries, most machines have some form of fluid logic incorporated within the control system. In construction equipment, hydraulic power is universally used and therefore offers many opportunities for utilizing fluid logic concepts and circuits for machine control—e.g., in vehicle propulsion on hydrostatic transmissions, in repetitive operations as on the Ford automatic backhoe, in safety accessories as on the anti-tilt devices on mobile cranes. Agricultural equipment represents one of the most lucrative areas for the application of fluid logic. Some of the areas which have already received considerable attention include:

> planters, pickers, tree shakers, hillside combines, constant depth and load plows, balers, windrowers, cultivators, feeders, potato harvesters, tree harvesters, feed handlers, and product elevators.

Other industries which are continually finding critical uses for fluid logic circuitry are associated with the following machines or systems:

INTRODUCTION

1. Automatic ore diggers, shovelers, and transporters.
2. Controllers for oil field leases—gaging and automatic custody.
3. Automatic web, film, or tape alignment, tension and rewinding controllers.
4. Automatic postal processing machines.
5. Automatic chemical processing systems.

1-5 GIST OF FLUID LOGIC DESIGN

Fluid logic offers a step forward in the concept of modern machine control. Tomorrow, and even right now, designers of automatic controls for machines will be expected to design complex logic control networks for the regimentation of manufacturing processes. Without a doubt, fluid logic provides another link in the chain of technological events which offer humanity the opportunity of eliminating drudgery and living in a new world only dreamed of a few years ago. It is the responsibility of those who profess to be designers of machines to expose themselves to the potential offered by fluid logic and consider its possibilities on future machine designs.

The area of fluid logic design encompasses the individual subjects of Boolean algebra, sensors to reflect operational states (the inputs of the logic system), motor circuits or the muscles of the machine, logic elements suitable for the control network, and synthesis procedures for deriving the optimal interconnection pattern of the logic network. Hence, to develop a fluid logic system, the designer needs a solid foundation in Boolean algebra, an awareness of hardware peculiarities, and a command of logic synthesis methodology. The logic design thrust vector depicts the facets of problem solution. See Figure 1-2.

Since the language of logic system design is Boolean algebra, its study cannot be ignored or deemphasized. Granted, little is gained by the designer overindulging in the heuristic aspects of the algebra; however, a familiarity with the postulates, theorems, and simplification techniques are indeed requisite to nontrivial logic problem solution. In this regard, and because this book is intended for the training of logic designers, the material presented on Boolean algebra is considered minimal and extremely compromising on time.

Because logic design acts as a bridge between system specification and hardware circuit implementation, the first step in the design process is to become acquainted with the problem or the machine application itself. This includes the establishment of the desired machine operation or cycle, the selection and description of the input signal generators for the control circuit, and the determination of the nature of the

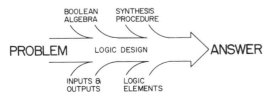

Figure 1-2 Logic design thrust vector.

network output signals needed to excite and control the motor systems of the machine. Only when such information is available can the actual synthesis of the logic system proceed.

The primary objectives when designing a fluid logic network are to attain low cost, high reliability, high speed, and a minimal number of elements that are easy to implement. To satisfy these objectives and to meet the logic requirements of the machine, the type of fluid elements which will be used to implement the network must be selected prior to initiating the synthesis. This requirement stems from the fact that optimal control networks are necessarily hardware-oriented and the type of logic elements selected for the system can severely influence the optimality of the final design.

For the designer, the most fascinating part of the logic system development is network synthesis. There are surprisingly few places in engineering and technology where the designer is fortunate to have a true algorithm for accomplishing a synthesis task. This indeed has been one of the hallmarks of logic design from its very inception. Since there are several approaches to logic synthesis, the designer is left with the enviable position of making his own procedural selection and accepting the best solution. A basis for making this selection is a major feature of this book, with computer-aided techniques included to enhance the designer's task. With the tools provided, the designer is armed with a rigorous methodology which will allow him to formulate a defendable position in support of his design.

DEFINITIONS

Air logic Relating to fluid logic components operated by air.
Circuit An assemblage of elements and their interconnection.
Combinational circuit A logic circuit in which the output at any given instant is dependent solely upon the present inputs.
Digital Relating to discrete operations or systems.
Element A device used to achieve a given function.
Fluidic Relating to nonmoving part fluid elements.
Fluid logic The study of a system utilizing digital control elements and circuits that uses fluids (liquids or gases) to transmit logic signals.
Hydra-logic Relating to fluid logic components operated by liquids.
Input A signal used to represent the state of a particular condition.
Logic Relating to reasoning and decision-making.
Network An assemblage of circuits and their interconnection.
Operation A defined action.
Output A signal used to represent the external state of a system.
Sensor An element which transforms a state property into a signal.
Sequential Occurring as a result of present and past states.
Sequential circuit A logic circuit in which the output at any given instant is dependent not only upon the present inputs but also on the input history.
Signal A detectable physical quantity or impulse representing the state of a system.
State A description of the condition, stage, position, level, etc., of a physical system.
System A collection of elements united by some form of interaction.

REFERENCES

Boole, George, *The Mathematical Analysis of Logic.* Cambridge, England, 1847. (Reprinted by Oxford, Basil, Blackwell of London in 1948).

Boole, George, *An Investigation of the Laws of Thought.* London, England, 1854. (Reprinted by Dover Publications, Inc., of New York in 1959).

Burchett, O'Neill J., "Application of Boolean Algebra to Hydraulic Circuit Simplification," M. S. Thesis, Oklahoma State University, 1960.

Cole, J. H., "Synthesis of Optimum Complex Fluid Logic Sequential Circuits," Ph.D. Dissertation, Oklahoma State University, 1968.

Fitch, E. C., "The Synthesis and Analysis of Fluid Control Networks," Ph.D. Dissertation, The University of Oklahoma, 1964.

Fitch, E. C., *Fluid Logic*, Teaching Manual Published by The School of Mechanical & Aerospace Engineering, Oklahoma State University, 1966.

Huffman, D. A., "The Synthesis of Sequential Switching Circuits," *Journal Franklin Institute*, Vol. 257, No. 3, pp. 161-190, March 1954.

Maroney, G. E., "A Synthesis Technique for Asynchronous Digital Control Networks," M. S. Thesis, Oklahoma State University, 1969.

Moore, E. A., "Gedanken—Experiments on Sequential Machines," *Automata Studies.* Princeton, New Jersey: Princeton University Press, pp. 129-153, 1956.

Ronan, H. R., "Hydraulic Switching Circuits," *Machine Design*, pp. 108-115, 140-145, February and March 1959.

Shannon, Claude E., "A Symbolic Analysis of Relay and Switching Circuits," *Trans. AIEE*, Vol. 57, pp. 713-723, December 1938.

Surjaatmadja, J. B., *A Computer-Oriented Method for Boolean Simplification and Potential Hazard Elimination*, Fluid Power Annual Research Conference, Report No. R73-FL-2, Fluid Power Research Center, Stillwater, Oklahoma, 1973.

Surjaatmadja, J. B., *A Generalized Method for Synthesizing Optimal Fluid Logic Networks*, Fluid Power Annual Research Conference, Report No. R73-FL-3, Fluid Power Research Center, Stillwater, Oklahoma, 1973.

Surjaatmadja, J. B., and Fitch, E. C., *Unique Machine States Through Output Considerations*, Fluid Power Annual Research Conference, Report No. P73-2, Fluid Power Research Center, Stillwater, Oklahoma, 1975.

Woods, R. L., "The State Matrix Method for the Synthesis of Digital Logic Systems," M. S. Thesis, Oklahoma State University, 1970.

Chapter 2

Boolean Algebra

2-1 THE ALGEBRA

The natural language used in everyday communication utilizes sentences (context) rather than mere words as the proper unit of meaning. It has evolved to such a refined stage that satisfactory understanding can be conveyed without the need for tiresome explanations. Unfortunately, the use of such vague and ambiguous information is totally unsuitable to convey ideas and instructions precisely enough for a machine. A machine is a perfect "listener" which is completely devoid of imagination, cannot guess, reacts literally, and responds at such a speed that correction of "misunderstanding" is impossible for long after it has reacted. A machine can never follow an unstated rule or an assumption. Thus, all operations must be presented or programmed in the form of detailed and precisely planned instructions.

To communicate conditions and relations in a simple and universally understandable form requires a language of exact thought. Symbolic logic satisfies this requirement both analytically and constructively. The science of propositional logic in symbolic form is termed "Boolean algebra." This language of symbolic logic is so powerfully structured that it is capable of making deductions and actually carrying out processes of reasoning. Fortunately, Boolean algebra is ideally suited as the

BOOLEAN ALGEBRA

language for logic design. Its brevity and its direct adaptation to practical problems make it particularly appealing to engineering design work.

Boolean algebra deals with variables having only two values, 0 and 1—hence the connotation binary. It involves a unique system of symbols and a precise and uncompromising set of rules governing the manipulation of the symbols. The symbols of the algebra are the letters assigned to the variables and the signs representing the operations. The rules are termed postulates and possess the axiomatic properties of coherence, contributiveness, consistency, and independence. These simple statements serve as the basis of the algebra and relate all the concepts involved.

Postulates of Boolean Algebra

$X = 1$ if $X \neq 0$ $\qquad$ $X = 0$ if $X \neq 1$
$1 \cdot 1 = (1 \text{ AND } 1) = 1$ $\qquad$ $0 + 0 = (0 \text{ OR } 0) = 0$
$1 \cdot 0 = (1 \text{ AND } 0) = 0$ $\qquad$ $0 + 1 = (0 \text{ OR } 1) = 1$
$0 \cdot 0 = (0 \text{ AND } 0) = 0$ $\qquad$ $1 + 1 = (1 \text{ OR } 1) = 1$
$1' = (\text{NOT } 1) = 0$ $\qquad$ $0' = (\text{NOT } 0) = 1$

Basically, the postulates define very succinctly the influence of the three basic operations—AND, OR, and NOT—upon the values of the variables. There is a symmetry in the postulates called duality, which means that for every proposition about sums there exists an analogous proposition about products, and vice versa. In other words, any law of the algebra that holds for addition holds also for multiplication. Thus, if products are associative, so are sums; and if sums are commutative, so are products; and so on. As a result of this duality, the postulates given above are presented in dual forms.

2-2 BOOLEAN THEOREMS

Unlike the postulates, Boolean theorems must contain nothing that cannot be proven, must be entirely implied by propositions other than themselves, and may contain no assumption not made in the postulates. Hence, the theorems are deduced entirely from the postulates and show the relationships among the implied concepts.

Boolean theorems are powerful propositions which are useful in establishing the identity of Boolean equations. They enable us to simplify logical expressions and transform them into other useful equivalent expressions. The implementation of Boolean network equations in terms of logic hardware is often an exercise in the manipulation and transformation of terms using Boolean theorems. A list of the theorems having particular significance in fluid logic is presented in Table 2-1.

Like the postulates, the theorems which are given in this chapter have been listed, where appropriate, as dual relations; for example, Theorem 9 is the dual of Theorem 10. The dual of a Boolean expression is obtained by changing all AND's to OR's, changing all 0's to 1's, and vice versa. Another form of a Boolean expression is the complementary expression. The complement of a Boolean expression is

Table 2-1 Boolean Theorems

The laws of tautology:	1	$X + X = X$
	2	$XX = X$
The laws of commutation:	3	$X + Y = Y + X$
	4	$XY = YX$
The laws of association:	5	$(X + Y) + Z = X + (Y + Z)$
	6	$(XY)Z = X(YZ)$
The laws of distribution:	7	$X + (YZ) = (X + Y)(X + Z)$
	8	$X(Y + Z) = (XY) + (XZ)$
The laws of absorption	9	$X + XY = X$
	10	$X(X + Y) = X$
The laws of the universe class:	11	$X \cdot 1 = X$
	12	$X + 1 = 1$
The laws of the null class:	13	$X \cdot 0 = 0$
	14	$X + 0 = X$
The laws of complementation:	15	$X + X' = 1$
	16	$XX' = 0$
The law of contraposition:	17	$X = Y' \Rightarrow Y = X'$
The law of double negation:	18	$X = X''$
The laws of expansion:	19	$XY + XY' = X$
	20	$(X + Y)(X + Y') = X$
The laws of DeMorgan:	21	$\overline{(X + Y)} = X'Y'$
	22	$\overline{XY} = X' + Y'$
The laws of reflection:	23	$X + X'Y = X + Y$
	24	$X(X' + Y) = XY$
The laws of transition:	25	$XY + YZ + X'Z = XY + X'Z$
	26	$(X + Y)(Y + Z)(X' + Z) = (X + Y)(X' + Z)$
The laws of transposition:	27	$XY + X'Z = (X + Z)(X' + Y)$
	28	$(X + Y)(X' + Z) = XZ + X'Y$

obtained in the same manner as the dual except that each literal is complemented. For example, consider the expression $XY' + XYZ'$—its complement is $(X' + Y)(X' + Y' + Z)$.

The various elements or letters comprising a Boolean equation are called the variables of the expression. Each individual entry of a variable into an equation is called a literal. Thus, an equation involving two variables may have any number of literals.

2-3 TRUTH TABLE PROOF

Although Boolean theorems can be proven by the use of other theorems, the most fundamental way of showing irrefutable evidence of a theorem's validity is by applying the truth table concept and the postulates of the algebra. This truth table proof consists of listing all combinations of values which the variables of the equivalence could possess and making substitutions of these values in both sides of the relation. If the value of both sides of the relation is the same after each substitution, then the equivalence is established.

BOOLEAN ALGEBRA

The use of the truth table method for proving a theorem is demonstrated with Theorem 23 as follows:

Variables X Y	Left Side $X + X'Y$	Right side $X + Y$
0 0	$0 + 1 \cdot 0 = 0$	$0 + 0 = 0$
0 1	$0 + 1 \cdot 1 = 1$	$0 + 1 = 1$
1 1	$1 + 0 \cdot 1 = 1$	$1 + 1 = 1$
1 0	$1 + 0 \cdot 0 = 1$	$1 + 0 = 1$

There are two variables involved in the theorem, and each variable can have only two explicit values—either 0 or 1; therefore, there are four combinations of values for the term XY. Substituting the values of the variables in the expressions for the left and right sides of the theorem and applying propositions expressed by the postulates, the values of each side can be determined. The theorem is valid if the value of the left side corresponds to the value of the right side for each combination of the values of the variables.

Each of the Boolean theorems presented in Table 2-1 can be proven by the use of the truth table technique. There is no better way to get acquainted with the postulates of Boolean algebra than to prove the validity of the theorems.

2.4 THE VENN DIAGRAM

One of the most effective ways to interpret and graphically represent the postulates, operators, and theorems of Boolean algebra is by the use of a Venn diagram. Such a diagram is a pictorial aid of the algebra where every condition of a system or class is symbolized by a specific location on the diagram. If a class K is defined as all possible regions within a square (the diagram), then any element A within the class is represented by a certain set of points within the square; e.g., all the points within a specific circle. Thus the element $(A \cdot B)$ can be defined as the largest region common to both A and B. Similarly, for the element $(A + B)$, the smallest region containing both A and B can uniquely describe the element. The Venn diagrams for these operations are illustrated in Figure 2-1. According to the postulates of the algebra, any system or class can have a value of either zero ($\phi = 0$) or one ($U = 1$).

The Venn diagram has been used to represent the Boolean operators and theorems and these are shown in Table 2-2. The reader should recognize that this

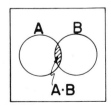
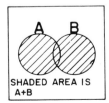

Figure 2-1 Venn diagrams for AND and OR.

Table 2-2 Descriptions of Boolean Operators and Theorems

CLASS.	LOGIC STATEMENT	LOGIC EQUATION	VENN representation	
			GEOMETRIC LOCATION	GEOMETRIC representation
BOOLEAN OPERATORS	OR	$X+Y$		
	AND	$X \cdot Y$		
	NOT	X'	$U=1$	
BOOLEAN THEOREMS	THE LAWS OF TAUTOLOGY	$X + X = X$		
		$X \cdot X = X$		
	THE LAWS OF COMMUTATION	$X+Y = Y+X$		
		$X \cdot Y = Y \cdot X$		
	THE LAWS OF ASSOCIATION	$(X+Y)+Z = X+(Y+Z)$		
		$(X \cdot Y) \cdot Z = X \cdot (Y \cdot Z)$		

CIRCUIT DIAGRAMS	TRUTH TABLES
X, Y → X+Y	X | Y | X+Y 0 | 0 | 0 0 | 1 | 1 1 | 1 | 1 1 | 0 | 1
X, Y → X·Y	X | Y | X·Y 0 | 0 | 0 0 | 1 | 0 1 | 1 | 1 1 | 0 | 0
X → N → X'	X | X' 0 | 1 1 | 0
X, X → X (OR) ; X, X → X (AND)	X | X+X | X·X 0 | 0 | 0 1 | 1 | 1
(X,Y)→ = (Y,X)→ ; (X,Y)→ = (Y,X)→	X | Y | X·Y | Y·X | X+Y | Y+X 0 | 0 | 0 | 0 | 0 | 0 0 | 1 | 0 | 0 | 1 | 1 1 | 1 | 1 | 1 | 1 | 1 1 | 0 | 0 | 0 | 1 | 1
associativity diagrams	X | Y | Z | (X+Y)+Z | X+(Y+Z) | (X·Y)·Z | X·(Y·Z) 0 | 0 | 0 | 0 | 0 | 0 | 0 0 | 0 | 1 | 1 | 1 | 0 | 0 0 | 1 | 1 | 1 | 1 | 0 | 0 0 | 1 | 0 | 1 | 1 | 0 | 0 1 | 1 | 0 | 1 | 1 | 0 | 0 1 | 1 | 1 | 1 | 1 | 1 | 1 1 | 0 | 1 | 1 | 1 | 0 | 0 1 | 0 | 0 | 1 | 1 | 0 | 0

Table 2-2 Descriptions of Boolean Operators and Theorems (*Continued*)

CLASS	LOGIC STATEMENT	LOGIC EQUATION	VENN representation	
			GEOMETRIC LOCATION	GEOMETRIC representation
BOOLEAN THEOREMS	THE LAWS OF DISTRIBUTION	$X+(Y \cdot Z) = (X+Y)(X+Z)$ $X \cdot (Y+Z) = (X \cdot Y)+(X \cdot Z)$		
	THE LAWS OF ABSORPTION	$X+(X \cdot Y) = X$ $X \cdot (X+Y) = X$		
	THE LAWS OF THE UNIVERSAL CLASS	$X \cdot 1 = X$ $X + 1 = 1$		
	THE LAWS OF THE NULL CLASS	$X \cdot 0 = 0$ $X + 0 = X$		

16

CIRCUIT DIAGRAMS	TRUST TABLES

X	Y	Z	X+(Y·Z)	(X+Y)(X+Z)	X·(Y+Z)	(X·Y)+(X·Z)
0	0	0	0	0	0	0
0	0	1	0	0	0	0
0	1	1	1	1	0	0
0	1	0	0	0	0	0
1	1	0	1	1	1	1
1	1	1	1	1	1	1
1	0	1	1	1	1	1
1	0	0	1	1	0	0

X	Y	X+(X·Y)	X·(X+Y)
0	0	0	0
0	1	0	0
1	1	1	1
1	0	1	1

X	1	X·1	X+1
0	1	0	1
1	1	1	1

X	0	X·0	X+0
0	0	0	0
1	0	0	1

17

Table 2-2 Descriptions of Boolean Operators and Theorems (*Continued*)

CLASS	LOGIC STATEMENT	LOGIC EQUATION	VENN representation	
			GEOMETRIC LOCATION	GEOMETRIC representation
BOOLEAN OPERATORS	THE LAWS OF COMPLEMENTATION	$X + X' = 1$		
		$X \cdot X' = 0$		
	THE LAWS OF CONTRAPOSITION	IF $X = Y'$ $\Rightarrow Y = X'$		
	THE LAWS OF DOUBLE NEGATION	$X'' = X$		
	THE LAWS OF EXPANSION	$X \cdot Y + X Y' = X$		
		$(X+Y) \cdot (X+Y') = X$		
	THE LAWS OF DeMORGAN	$\overline{X+Y} = X' \cdot Y'$		
		$\overline{X \cdot Y} = X' + Y'$		

CIRCUIT DIAGRAMS	TRUTH TABLES

0	1	X	X'	X+X'	X·X'
0	1	0	1	1	0
0	1	1	0	1	0

X	Y	X'	Y'
0	1	1	0
1	0	0	1

X	X'	X''
0	1	0
1	0	1

X	Y	X·Y+X·Y'	(X+Y)·(X+Y')
0	0	0	0
0	1	0	0
1	1	1	1
1	0	1	1

X	Y	$\overline{X+Y}$	$\overline{X \cdot Y}$	X'·Y'	X'+Y'
0	0	1	1	1	1
0	1	0	1	0	1
1	1	0	0	0	0
1	0	0	1	0	1

Table 2-2 Descriptions of Boolean Operators and Theorems (*Continued*)

CLASS	LOGIC STATEMENT	LOGIC EQUATION	VENN representation	
			GEOMETRIC LOCATION	GEOMETRIC representation
BOOLEAN THEOREMS	THE LAWS OF REFLECTION	$X + X' \cdot Y = X + Y$		
		$X \cdot (X' + Y) = X \cdot Y$		
	THE LAWS OF TRANSITION	$X \cdot Y + Y \cdot Z + X' \cdot Z$ $= X \cdot Y + X' Z$		
		$(X+Y) \cdot (Y+Z)$ $\cdot (X'+Z) = (X+Y) \cdot (X'+Z)$		
	THE LAWS OF TRANSPOSITION	$X \cdot Y + X' \cdot Z =$ $(X+Z) \cdot (X'+Y)$		
		$(X+Y) \cdot (X'+Z) =$ $X \cdot Z + X' \cdot Y$		

graphical means of representation becomes too impractical for complex Boolean equations.

2-5 CANONICAL FORMS

Boolean equations can contain both individual literals or combinations of literals called terms (a product term or a sum term). A product term is a group of two or more variables in complemented or uncomplemented form which are interconnected

BOOLEAN ALGEBRA

| CIRCUIT DIAGRAMS | TRUTH TABLES |

X	Y	X+X'·Y	X·(X'+Y)	X+Y	X·Y
0	0	0	0	0	0
0	1	1	0	1	0
1	1	1	1	1	1
1	0	1	0	1	0

X	Y	Z	X·Y+X'·Z+Y·Z	X·Y+X'·Z	(X+Y)·(X'+Z)·(Y+Z)	(X+Y)·(X'+Z)
0	0	0	0	0	0	0
0	0	1	1	1	0	0
0	1	1	1	1	1	1
0	1	0	0	0	1	1
1	1	0	1	1	0	0
1	1	1	1	1	1	1
1	0	1	0	0	1	1
1	0	0	0	0	0	0

X	Y	Z	X·Y	X'·Z	X+Z	X'+Y	XY+X'·Z	(X+Z)·(X'+Y)
0	0	0	0	0	0	1	0	0
0	0	1	0	1	1	1	1	1
0	1	1	0	1	1	1	1	1
0	1	0	0	0	0	1	0	0
1	1	0	1	0	1	1	1	1
1	1	1	1	0	1	1	1	1
1	0	1	0	0	1	0	0	0
1	0	0	0	0	1	0	0	0

by AND's. A sum term, on the other hand, is a group of two or more variables in complemented or uncomplemented form that are interconnected by OR's. The exclusive use of either term in an expression provides a unique means of categorizing these equations.

There are two standard or normal forms in which Boolean equations can be written. These forms are the disjunctive form and the conjunctive form. A disjunctive expression contains either a single product term or a sum of product terms. A conjunctive expression possesses either a single sum term or a product of sum terms.

Examples of disjunctive equations are:

$X = AB$, $X = AB + CD$, and $X = A + BC + DEF$

Typical conjunctive-type equations are:

$X = A + B$, $X = (A + B)(C + D)$, and $X = A(B + C)(D + E + F)$

If a product term contains every variable represented in a given expression in either its complemented or uncomplemented form, it is called a "minterm." Similarly, if a sum term contains every variable represented in an associated expression in either its complemented or uncomplemented form, it is called a "maxterm." A disjunction of minterms is an important expanded form of an expression and consists of only minterms—it is called the disjunctive canonical form. Likewise, a conjunction of maxterms is an expanded form consisting of only maxterms—it is called the conjunctive canonical form.

Both the expanded disjunctive and conjunctive forms can be derived for any relationship by using its truth table representation. To illustrate these derivations, consider the following expression:

$$X = AB + AC' + B'C \tag{2-1}$$

The truth table would be:

A	B	C	X
0	0	0	0
0	0	1	1
0	1	1	0
0	1	0	0
1	0	0	1
1	0	1	1
1	1	1	1
1	1	0	1

The expanded disjunctive expression is written by including the appropriate term where the value of X is equal to one. Hence, the disjunction of minterms is:

$$X = A'B'C + AB'C' + AB'C + ABC' + ABC \tag{2-2}$$

To derive the expanded conjunctive expression, the complemented disjunctive equation is first written by including the appropriate term where the value of X is equal to zero. Hence, the complementation yields the following disjunction of minterms:

$$X' = A'B'C' + A'BC + A'BC' \tag{2-3}$$

BOOLEAN ALGEBRA

The conjunction of maxterms is obtained by complementing Equation 2-3, which yields:

$$X = (A + B + C)(A + B' + C')(A + B' + C) \qquad (2\text{-}4)$$

Applying Theorem 15 yields still another way of obtaining minterms and maxterms for achieving the expanded canonical forms of expressions. To demonstrate this, consider Equation 2-1 and apply Theorem 15 as follows:

$$X = AB(C + C') + A(B + B')C' + (A + A')B'C$$

Applying Theorem 8 yields the same disjunction of minterms given in Equation 2-2.

A conjunctive expression can be transformed into expanded canonical form by a process similar to that used for the disjunctive case. Consider the following conjunction:

$$X = (A + C)(A + B') \qquad (2\text{-}5)$$

Theorem 16 can be applied as follows:

$$X = ((A + C) + BB')((A + B') + CC')$$

Utilizing Theorem 7 appropriately yields the necessary maxterms for the expanded conjunction in Equation 2-6.

$$X = (A + B + C)(A + B' + C)(A + B' + C)(A + B' + C') \qquad (2\text{-}6)$$

2-6 UNCOMPLEMENTATION ALGORITHM

Complex complemented forms of Boolean equations generally arise when circuits involving either NOR or NAND logic elements are derived or analyzed. The resolution of these equations into uncomplemented forms constitutes an important facet of fluid logic. Fortunately, the repetitive application of DeMorgan's law of complementation offers an effective means of reducing such complex types of equations to an uncomplemented and manageable form.

To implement the use of DeMorgan's Theorem (Theorems 21 and 22), the following algorithm has been developed:

1. Bracket all groups of literals under each complementation line.
2. Count the number of complementations above each literal and each operator (AND and OR), and:
   a. If the number of complementations acting on a literal is even, retain the uncomplemented literal
   b. If the number of complementations above an operator is even, retain the same operator

    c If the number of complementations acting on a literal is odd, use the complemented literal

    d If the number of complementations above an AND operator is odd, replace the AND with an OR

    e If the number of complementations above an OR operator is odd, replace the OR with a reverse bracketed AND, i.e.,) · (.

3 Respecting all brackets, both original and added, record the solution simultaneously while performing Step 2.

4 Eliminate any redundant brackets.

In order to illustrate the use of the "uncomplementation algorithm," consider the following problem:

$$F = \overline{\overline{\overline{ab' + a'b} + ac'} + \overline{a' \ (cd' + \overline{(b + d')})}} \tag{2-7}$$

Performing Step 1 of the algorithm yields:

$$F = \overline{\overline{((\overline{ab' + a'b}) + ac')} + \overline{(a' \ (cd' + \overline{(b + d')}))}}$$

Performing Steps 2 and 3 results in the following:

$$F = (((a' + b) \cdot (a + b') + a \cdot c')) \cdot ((a' \ (c \cdot d' + (b') \cdot (d))))$$

By eliminating the redundant brackets, the following solution results:

$$F = ((a' + b)(a + b') + ac')(a' \ (cd' + b'd)) \tag{2-8}$$

2-7 KARNAUGH MAPS

Using Boolean equations as the basis for implementing fluid logic networks requires that the equations exhibit near minimal form. When the number of variables involved in an expression is small (preferably six or less), one of the most powerful techniques for achieving simplification is the Karnaugh map method. This simple and rapid technique provides a unique means for obtaining a minimal sum or product—either a disjunction or a conjunction containing the fewest possible number of terms and/or literals.

    The Karnaugh map provides a graphical means of representing a Boolean equation and allows a visual-type inspection procedure to be applied in achieving simplification. This basic method for describing and manipulating logic specifications is so important in understanding the implications of logic interrelationships that the subject deserves special attention.

    A Karnaugh map must contain a sufficient number of rows and columns to provide appropriately a map location or cell for each of the 2^n possible combinations of variable values. The map uses the reflected binary (Gray Code) code for labeling the rows and columns, because in this binary numbering system only one

BOOLEAN ALGEBRA

Table 2-3 The Reflected Binary (Gray) Code

0	0	0
0	0	1
0	1	1
0	1	0
1	1	0
1	1	1
1	0	1
1	0	0

binary bit changes at a time. The Gray Code is shown in Table 2-3 and can be extended indefinitely as long as one binary bit changes at a time.

The blocks composing the map created by the intersections of the rows and columns are called "cells" in which each minterm or maxterm of a disjunctive or conjunctive canonical equation can be identified. Hence, a Karnaugh map is a graphical representation of a fully expanded disjunctive or conjunctive equation much like truth tables. However, a map, in contrast to the truth table, exhibits basic patterns which permit the simplest expression to be read directly.

The basic patterns displayed by cells containing entries are called "subcubes." More specifically, a subcube is a set of cells representing all or a portion of the minterms or maxterms of a Boolean equation which possesses an adjacency relationship. Such a relationship occurs when one or more of the variables defining the cells of the subcube have constant values (common variable values).

A Karnaugh map is applicable to Boolean equations having disjunctive or conjunctive form. Equations involving the cells with "ones" provide the uncomplemented solution, while equations involving only zeros yield the complemented solution. When dealing with disjunctive equations, a given equation is satisfied (equal to 1) when the values of the variables comply with the requirements of any product (AND) term in the equation. A specific entry in a cell of the map prescribes the combination of variables needed to give one solution to the equation. Hence, a value of one is entered in a cell that describes the proper values of the variables needed by a particular product term to yield a solution. To illustrate a two-variable map, consider the following expression:

$$Z = AB + A'B \tag{2-9}$$

This equation can be satisfied ($Z = 1$) by two different variable combinations—either $A = 1$ AND $B = 1$ or $A = 0$ AND $B = 1$. Also since this expression is a two-variable relation and both terms are already in minterm form, the map will contain two entries. The map must have 2^2 or four locations or cells. The Karnaugh map representing Equation 2-9 is shown in Figure 2-2. Note that a 1 is entered in the cell or cells describing each minterm of the equation. Therefore, for the term AB, a 1 is entered in the cell where $A = 1$ and $B = 1$. Likewise for the term $A'B$, a 1 is recorded in the cell where $A = 0$ and $B = 1$.

Figure 2-2 Karnaugh map for Equation 2-9.

Two cells possess an adjacency relationship if the values of the variables which describe each cell are dissimilar in only one variable. The two cells containing 1's in Figure 2-2 are adjacent, since only the value of the variable A changes in going from one cell to the other. For this reason the two cells constitute a subcube, which can be described by only a single variable; hence, the simplest expression for Equation 2-9 is merely $Z = B$.

In the case of a three-variable equation such as:

$$A = XY + YZ + X'Z \qquad (2\text{-}10)$$

the Karnaugh map must contain 2^3 or eight cells. The appropriate entries for the three terms are shown in the eight-cell map of Figure 2-3. Since the terms in the equation are not minterms, two entries are required in the map to represent one term. Note that there are three possible subcubes reflected by the map—$X'Z$, YZ, and XY. Since a given cell needs to be covered only once by a subcube, the simplest expression to represent Equation 2-10 is:

$$A = XY + X'Z \qquad (2\text{-}11)$$

The three (two-cell) subcubes used to describe Equation 2-10 contain the redundant subcube YZ. However, it should be stated that overlapping or "chained subcubes" are always permissible and their use is encouraged particularly when they assist in yielding a simpler relation.

It should become obvious that the simplest equations are normally obtained by using the largest subcubes. The terms represented by these subcubes are the prime implicants for a disjunctive expression and the prime implicates when the Karnaugh map is constructed for a conjunctive expression. These "prime" terms are the irreducible terms of a Boolean equation. The expression required to cover a single cell must include every variable on the map, whereas a two-cell subcube eliminates one of the variables, a four-cell subcube reduces the describing variables by two and an eight-cell subcube eliminates three variables. This subcube concept for eliminating the number of describing variables can be demonstrated by Figure 2-4. The equation representing the map requires a four-variable, a three-variable, and a two-variable product term to include the single cell, the two-cell subcube, and the four-cell subcube, respectively. The equation reflected by Figure 2-4 is:

Figure 2-3 Karnaugh map for Equation 2-10.

BOOLEAN ALGEBRA

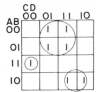

Figure 2-4 Karnaugh map for Equation 2-12.

$$Y = ABC'D' + AB'C + A'D \qquad (2\text{-}12)$$

Although the Karnaugh map completely describes a Boolean equation in all of its forms, a reasonable amount of practice is required to recognize optimum subcubes to yield minimal relations. The adjacency characteristics of peripheral cells are particularly difficult to recognize in subcube patterns. To illustrate subcube identification, various examples of three- and four-variable maps are presented in Figures 2-5 and 2-6. It is very important that the three- and four-variable maps be completely mastered before attempting to read larger maps.

It has been mentioned that the goal in reading a Karnaugh map is to recognize optimum subcubes in order to write minimal equations. In many cases, there are several subcube combinations which will yield equations having the same number of literals or connectives. In order to illustrate multiple minimal relations, consider the following equation:

$$Z = A'C' + AD + BCD' + A'BCD + AB'CD' \qquad (2\text{-}13)$$

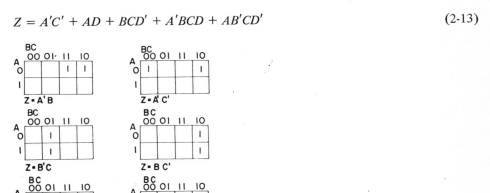

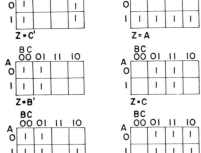

Figure 2-5 Examples of subcubes on three-variable maps.

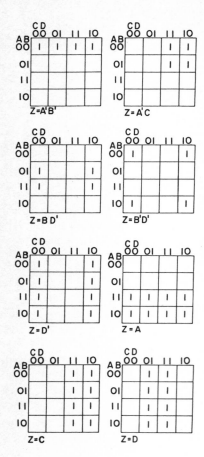

Figure 2-6 Examples of subcubes on four-variable maps.

The Karnaugh map which completely describes Equation 2-13 is shown in Figure 2-7. Two of the minimal equations (eight literals) which can be written from the map are:

$$Z = A'C' + C'D + A'B + AC \qquad (2\text{-}14)$$
$$Z = A'C' + C'D + AC + BC \qquad (2\text{-}15)$$

In practice, both solutions should be investigated in order to determine the one which can be best implemented with hardware.

Figure 2-7 Karnaugh map for Equation 2-13.

BOOLEAN ALGEBRA

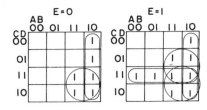

Figure 2-8 A five-variable map for Equation 2-16.

The size of a Karnaugh map doubles for each variable added. In order to illustrate the use of maps for relatively large numbers of variables and also to show the complexity involved, examples will be given involving five, six, seven, and eight variables. A five-variable map pattern is shown in Figure 2-8 for the relation:

$$Z = ABC'DE + ABC + AB' + CDE \qquad (2\text{-}16)$$

One group of optimum subcubes gives the simplified equation as:

$$Z = AC + AB' + CDE + ADE \qquad (2\text{-}17)$$

A six-variable map is given in Figure 2-9 for the relation:

$$Z = AB'CDE + ABE + CD'E + E'F + CD'E' \qquad (2\text{-}18)$$

The reader should try to write a simplified equation from the map and compare his minimal form with the following solution:

$$Z = E'F + CD' + ACE + ABE \qquad (2\text{-}19)$$

A seven-variable Karnaugh map is given in Figure 2-10 for the relation:

$$Z = A'BE' + ACE'F'G + DE'F'G' + A'BE + DEF'G' \qquad (2\text{-}20)$$

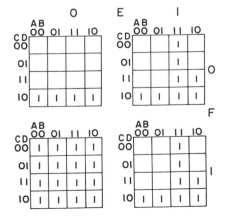

Figure 2-9 A six-variable map for Equation 2-18.

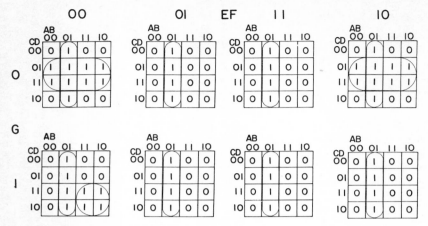

Figure 2-10 A seven-variable map for Equation 2-20.

The simplified function is:

$$Z = A'B + DF'G' + ACE'F'G \qquad (2\text{-}21)$$

Finally, an eight-variable map is shown in Figure 2-11 which represents the equation:

$$Z = A'G'H' + A'C'H' + ACG' + A'C'GH + A'C'G' \qquad (2\text{-}22)$$

The simplified equation is:

$$Z = A'C' + CG'H' + ACG' \qquad (2\text{-}23)$$

The rules for making entries in the maps have considered only solution entries or 1 entries. The remaining entries in the map are of two types: complementary entries or "don't care" entries. The complementary entries are 0 entries and are recorded in locations where operations or solutions must not occur. The "don't care" or optional entries are − entries where the values of the variables never occur or if they do occur the result would not influence the logic of the system.

The optional or − entries can be utilized in simplifying the switching equation by including them when necessary to complete optimum subcubes. To illustrate the use of − entries, consider the expression:

$$Z = A'B'D' + B'C'D' + A'BC'D + ABCD \qquad (2\text{-}24)$$

and the "don't care" entries $A'BCD$, $ABC'D$, and $AB'CD'$. The Karnaugh map for Equation 2-24, including complementary and optional entries, is given in Figure 2-12. Using the optional entries as desired to complete the optimum subcubes yields

BOOLEAN ALGEBRA

the following simplified solution:

$$Z = B'D' + BD \qquad (2\text{-}25)$$

Synthesis maps for logic circuits generally contain many solution entries 1's or many complementary entries 0's. When the map displays only a few 0 entries, it may prove simpler to express the complementary solution. This solution considers all the 0 entries rather than the 1 entries. The uncomplemented solution is derived by complementing the "zero" group expression. The use of the complementary solution is illustrated by considering the following expression:

$$Z = BD + AD' + AB'CD \qquad (2\text{-}26)$$

The Karnaugh map for Equation 2-26 is presented in Figure 2-13. The

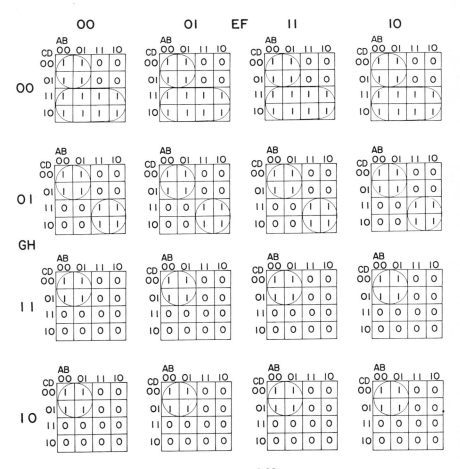

Figure 2-11 An eight-variable map for Equation 2-22.

Figure 2-12 Use of optional entries.

complementary solution is:

$$Z' = A'B' + A'D' \tag{2-27}$$

Taking the complement yields:

$$Z = (A + B)(A + D) \tag{2-28}$$

Using Theorem 7 gives:

$$Z = A + BD \tag{2-29}$$

which in this case could have been read directly from the map by using the solution entries.

Karnaugh maps can also be utilized for the manipulation of conjunctive Boolean expressions. For example, consider the following conjunctive equation:

$$Z = (A' + B' + C' + D')(A' + C' + D)(B + C + D)(A + C + D)(A' + B' + D) \tag{2-30}$$

In this case, the sum term $(A' + B' + C' + D')$ is entered in the map cell where the values of the variables $A, B, C,$ and D are all zero. Likewise, the term $(A' + C' + D)$ is entered in the two appropriate cells, satisfying the variable values of $A = C = 0$ and $D = 1$. If the remaining terms are entered in the same way, the completed Karnaugh map will be as shown in Figure 2-14. By selecting the three subcubes as shown in Figure 2-14 to represent the implied logic, the simplest conjunctive equation can be read directly from the map as:

$$Z = (A' + B' + C')(A' + D)(C + D) \tag{2-31}$$

AB\CD	00	01	11	10
00	0	0	0	–
01	0	1	1	0
11	1	1	1	1
10	1	–	1	1

Figure 2-13 Use of complementary entries.

BOOLEAN ALGEBRA

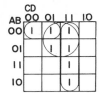

Figure 2-14 Karnaugh map for Equations 2-30, 2-31, and 2-32.

Note that the terms read from the Karnaugh map are read in the conjunctive form, except that the "universe" is zero—a key factor in the application of this technique.

The validity of the conjunctive approach can be demonstrated simply by using the principle of duality—showing that the above forms are equivalent. Consider the following disjunctive equation:

$$Z = A'B'C'D' + A'C'D + BCD + ACD + A'B'D \qquad (2\text{-}32)$$

The Karnaugh map for this equation will be identical to the map for Equation 2-30. Hence, it can be derived directly from the map of Figure 2-14, i.e.:

$$A'B'C'D' + A'C'D + BCD + ACD + A'B'D = A'B'C' + A'D + CD$$

Since the equation is the dual of Equations 2-30 and 2-31 by inspection, these equations are equivalent, and for every conjunction equation a relating disjunctive dual can be constructed. This proves the validity of the approach.

2-8 COMPUTER-AIDED SIMPLIFICATION

As has already been indicated, Karnaugh maps, like all other manual/graphical techniques, are not suitable for simplifying Boolean equations containing more than seven or eight variables. This is due mainly to the fact that the size of the problem actually doubles for every variable that is added to an equation. The need to consider problems having a great number of variables stems from the advantages offered by complex sequential networks as well as from sophisticated combinational logic circuits.

It was recognized by early investigators, e.g., W. V. Quine and E. J. McCluskey, Jr., that the only hope for the simplification of large variable equations was in the use of computerized techniques. To this end, they advanced a numerical-type tabular technique which was designed to serve as a starting point for those that followed. Although many different attempts have been made to develop a suitable computerized simplification method, the one described here is the only method known to the authors which will satisfy the requirements in fluid logic. Full details of the simplification method are contained in the Surjaatmadja references at the end of this chapter.

The referenced program is called TAB II, is written in Fortran IV, and is based on a special tabular technique which does not require any expansion of the original expression. The maximum size of the equation which can be handled is limited only by the size of the computer. The authors have personally used the program for equations involving up to eighty variables and thirty terms. TAB II will accept an equation in either disjunctive or conjunctive forms, but will always give the resulting answer in the same form. The program has been designed to accept both the "don't care" terms as well as the required "off" terms of a circuit problem. TAB II automatically detects any insufficiency of the computer to handle a problem and gives a message to help the user in knowing which step the computer cannot execute.

In order to provide the reader with some detailed insight as to the nature of TAB II, a brief operational description is presented. The input to the program contains the following:

1. Number of optional terms.
2. Number of primary terms.
3. Number of variables involved.
4. Function variable indicating the type of output required.
5. Option variable, relating to the required procedure to obtain the "don't cares."
6. Actual terms of the equation.

To illustrate the format used for the input and output, consider the simplification of the following two Boolean expressions (one in disjunctive form and the other in conjunctive form):

1. $abc'defgh' + ab'dg'h + abc'dg'h' + abcdefh' + bdg'h + bdgh$
2. $(a + b + c' + d + e + f + g + h')(a + b' + d + g' + h)$
   $(a + b + c' + d + g' + h')(a + b + c + d + e + f + h')$
   $(b + d + g' + h)(b + d + g + h)$

For these example problems, the actual input and outputs of the computer are the same. The input deck is represented as follows:

Card	Entry
	1 2 3 4 5 6 7 8 9 0 1 2 3 4 5 6 7 8 9 0 1 2 3 4
1	0 6 8 0 0 8
2	1 1 0 1 1 1 1 0
3	1 0 2 1 2 2 0 1
4	1 1 0 1 2 2 0 0
5	1 1 1 1 1 1 2 0
6	2 1 2 1 2 2 0 1
7	2 1 2 1 2 2 1 1

where 2's are used to represent indeterminate variable values, indicating that the corresponding term is not dependent upon the particular variable. The computer output is illustrated as follows:

BOOLEAN ALGEBRA

Problem Statement
The Primary Terms:
Variable No.:

0	0	0	0	0	0	0	0
1	2	3	4	5	6	7	8

Term no:

1	1	1	0	1	1	1	1	0
2	1	0	2	1	2	2	0	1
3	1	1	0	1	2	2	0	0
4	1	1	1	1	1	1	2	0
5	2	1	2	1	2	2	0	1
6	2	1	2	1	2	2	1	1

The minimized solution is requested.

Problem Solution:
The Minimized Representation:
Variable No.:

0	0	0	0	0	0	0	0
1	2	3	4	5	6	7	8

Term no.:

1	1	1	0	1	2	2	0	2
2	1	1	2	1	1	1	2	2
3	1	2	2	1	2	2	0	1
4	2	1	2	1	2	2	2	1

The computer solutions can be interpreted as:

$abc'dg' + abdef + adg'h + bdh$

or, for a conjunctive expression:

$(a + b + c' + d + g')(a + b + d + e + f)(a + d + g' + h)(b + d + h)$

PROBLEMS

1. Prove by using the truth table method Theorems 1 through 10. Use only the postulates of Boolean algebra as the basis of proof.
2. Using any method desired, prove the validity of Theorems 23 through 28.
3. Represent the following equations in the Venn diagram:
   a. $Z = AB + B'C + A'B'$
   b. $Z = A'B'C' + A'C$
   c. $Z = B(A'(C + BD')(A' + CD) + CD')$
4. Show the dual and the complement of the following expressions:
   a. $AB + A'C + B'C'$
   b. $AD + BC' + A'X + C'D$

   c $A B D' + A'C + A B'C D + B C' D$
   d $Z Y + Y' Z + W Z'$

5 Obtain the canonical forms of the following equations by the truth table approach and the theorem approach:
   a $Z = a'b'c + ab'd + cd'e + abce'$
   b $Z = pr' + pst'x + p't'x' + rs't + tx$
   c $Z = (a' + b + c')(b + d' + e)(a + b')(a' + d')$
   d $Z = dc(a + e')(b + d' + e')(a' + b)$

6 By the uncomplementation algorithm, simplify the following expressions:
   a $\overline{\overline{(A + B)} + \overline{(A' + B + C')}}$
   b $\overline{\overline{A B'} + \overline{A' B} + A C' (B + D')}$
   c $\overline{A' C + A B C' + B' C + A'(C D' + B)}$
   d $[AB'(\overline{CD' + E}) + FG][(A'D + \overline{CF'} + B'G) + E]$
   e $[D'(AC' + \overline{BD'E}) + AE'\overline{(B'D + F)} + \overline{CF}]$

7 Simplify the following expressions:
   a $AB(A' + B C')$
   b $A(B + CD + C' + D') + E$
   c $A' B C' + A' B C + A B' C + \underline{A B C' + A B C}$
   d $(A'B + AB') + (A'B' + A B)[(ACD + B) + B C' + A'D']$
   e $WX + XY + X'Z' + WY'Z$
   f $W'X' + X'Y' + YZ + W'Z'$
   g $(A + BC)\{D'E'[C' + A'(B + C)]\} + (B' + D')(B + D) + B'C'DE$

8 Determine the validity of the following equations:
   a $A'B'D' + B C + A B'C = \underline{B'C'D' + A B'D + A'C D' + A B C + B C D}$
   b $A' B' + B C + B' C' = B'(A B' C) + A B C + A'C$
   c $B'C' + A C' + A B = A'B'C' + A B'C' + A B$
   d $A'B'C' + A' B D + A B C + A B'D' = B'C'D' + A'C'D + B C D + A C D'$
   e $A'C' + B C = A'B'C' + A'B + A B C = A' B D + A'B'C' + A'BD' + A B C$
   f $C'(A + B)(A'B') + A C(A' + B) + A'C(A + B') = A B' C' + A B C + A' B C' + A' B' C D + A' B' C' D'$
   g $A'C'D' + BD + AD + ABC =$
      $(A' + C + D)(A' + B + C' + D)(A + B + D')(A + B + C')(A + C' + D)$
   h $A'B'C' + AC' + A'BC + BC' = (A + B + C')(A' + B' + C')$

9 Using the Karnaugh maps, obtain the most simplified form of the following expressions:
   a $X = B'C' + A B D' + A B' C + B'C D + B C D' + A'B'C$
   b $X = A'C'D' + B'D + A'C D' + A B D' + A B C$
   c $X = A'B'D + A'B C' + A B C' + A C$
   d $X = B'C'D' + A'B C' + A B C'D + A'B' C D + A B' C D + A'B C D'$
   e $X = A' B' D + A B C' + B D + B C D'$
   f $X = A' C' + A C' D' + A C D'$
   g $X = A'B'D' + A B C' + A B' C D E' + A B' C E + A' B C D E' + A' B C D' E' + A' B C D E$
   h $X = A B'F' + C D E + A B C D'E + A B' E F + A' E' F + A C'E'F$
   i $X = (A + B + C)(A' + B' + C')(A + B' + C)(A' + B + C')$
   j $X = (B' + C + D)(A + B + D')(B' + C' + D)(A' + B + D')$
   k $X = (A + B)(B' + C')(A' + D)(A + C')(B' + D)$

BOOLEAN ALGEBRA

10 Obtain the most simplified conjunctive form of the following disjunctive expressions:
   a $A + B'CD' + AD' + BC + A'C$
   b $A'BC' + A + B' + D + A'B'D + E'$
   c $A'B'C'D' + AB'C'D' + A'B'CD' + AB'CD' + A'BCD + ABCD$
   d $A'BC'D' + A'B'C'D + A'BCD' + AB'C'D' + AB'CD' + ABC'D'$
   e $AB'C'D + ABC'D' + ABC'D + ABCD' + ABCD$

11 Obtain the simplified disjunctive form of the following conjunctive expressions:
   a $(A + C')(B + C')(A' + B')$
   b $(B + D')(C + D)(A + B')$
   c $(A + B')(B + C')(C + D')(A' + D)$
   d $(A + B)(B + C)(A + A)$
   e $(A' + B')(B' + C)(C' + A')$

12 Write the simplest disjunctive and conjunctive Boolean expressions represented by the Karnaugh map shown in Figure P2-12. Assume that the solution entries represent a disjunctive expression.

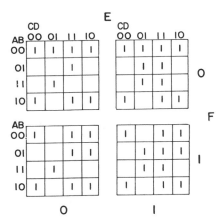

Figure P2-12

13 Using the Karnaugh map, obtain the most simplified form of the following equations. Make use of the available "don't care" terms; and when the "OFF" terms are given, assume that the remaining subcubes of the map are "don't care" locations. Also obtain the most simplified solution in the conjunctive form if a disjunctive form is given, and vice versa.
   a $Z = ab'c + bcd' + abd$
      $Z_{OFF} = ab'c' + a'c'd'$
   b $Z = ab'cd + a'cde + b'c'de$
      $Z_{d.c.} = a'c'de' + bc'de + abcde$ (Subscript d.c. denotes "don't cares.")
   c $Z = ab'f' + cde + abcd'e + ab'ef + a'e'f + ac'e'f$
      $Z_{d.c.} = abc'f' + abce'f' + abc'ef$
   d $Z = ab'f' + cde + abcd'e + ab'ef + a'e'f + ac'e'f$
      $Z_{OFF} = a'b'c'e'f' + a'cd'ef' + a'c'd'ef$
   e $Z = (a + b' + c)(a' + c' + d + e')(a' + b' + e)(b + c' + d + e)$
      $Z_{d.c.} = (a' + b + c + d)(b + c' + d' + e)(a + b' + c' + e)$
      $(a + b' + c' + d)$

$$\mathbf{f} \quad Z = (b' + c' + e + f)(a + b' + c' + e')(a' + b + c' + f)(a + c' + d + e)$$
$$Z_{\text{OFF}} = a'cd'e' + a'b'cdef'$$

DEFINITIONS

Class A group sharing common conditions.

Complement The mathematical inverse of the entity. Two quantities or symbols are complements of each other if and only if one quantity or symbol takes the value of one when the other equals zero, and vice versa.

Complementary expression An expression whose literals, operators, and constants have inverted forms, i.e., 0, 1, +, ·, complemented literals and uncomplemented literals in one expression correlates with 1, 0, ·, +, uncomplemented literals, and complemented literals in its complementary expression.

Conjunctive canonical form A conjunction of maxterms.

Conjunctive expression A Boolean expression that contains a single sum term or a product (conjunction) of sum terms.

Conjunctive form A product (conjunction) of sum terms.

Constant A symbol that represents the values of the algebra and can be only 0 or 1.

Disjunctive canonical form A disjunction of minterms.

Disjunctive expression A Boolean expression that contains a single product term or a sum (disjunction) of product terms.

Disjunctive form A sum (disjunction) of product terms.

Don't care terms Minterms and maxterms which can be added to a function without affecting its desired operation.

Dual expressions Expressions whose operators and constants have inverted forms, i.e., 0, 1, +, and · in one expression correlates with 1, 0, ·, and + in its dual, respectively.

Equation An equality involving two or more constants, variables, or functions.

Equivalent expressions Expressions which have the same value when the values of the variables are equal.

Expression A combination of variables and operators.

Function A mathematical expression describing the relation between variables.

Implicant of a function A product term of a function that does not contain any pair of complementary literals (e.g., aa').

Implicate of a function A sum term of a function that does not contain any pair of complementary literals (e.g., $a + a'$).

Indeterminate variable value A variable value that can be any of the two values of the algebra.

Literal An individual entry of a variable into an equation in either its complemented or its uncomplemented form.

Maxterm A sum term that contains every variable in the system in either its complemented or its uncomplemented form.

Minterm A product term that contains every variable in the system in either its complemented or its uncomplemented form.

Normal form A disjunctive form or a conjunctive form.

Operator A symbol that denotes a mathematical operation.

Prime implicant of a function An irreducible implicant of a function.

Prime implicate of a function An irreducible implicate of a function.

Product term A conjunction of literals, a group of literals interconnected by AND's.
Sum term A disjunction of literals, a group of literals interconnected by OR's.
Variable The various letters and/or symbols that are not operators.
Variable value A constant that represents a numerical condition of a variable.

REFERENCES

Dietmeyer, D. L., *Logic Design of Digital Systems*. Boston: Allyn and Bacon Inc. 1971.

Fitch, E. C., Jr., *Fluid Logic*, Teaching Manual Published by The School of Mechanical and Aerospace Engineering, Oklahoma State University, 1966.

Hilton, A. M., "Logic and Switching Circuits," *Electrical Manufacturing*, Vol. 65, No. 4, pp. 123-158, April 1960.

Hohn, F. E., *Applied Boolean Algebra—An Elementary Introduction*. New York: The Macmillan Company, 1966.

Humphrey, W. S., Jr., *Switching Circuits*. New York: McGraw-Hill Book Company, 1958.

Karnaugh, M., "The Map Method for Logic Synthesis of Combinational Logic Circuits," *AIEE Trans. on Communications and Electronics*, Vol. 72, pp. 593-599, Nov. 1953.

McCluskey, E. J., Jr., "Minimization of Boolean Functions," *Bell Systems Technical Journal*, Vol. 25, pp. 1417-1444, Nov. 1956.

Quine, W. V., "The Problem of Simplifying Truth Functions," *American Math. Monthly*, Vol. 59, pp. 521-531, Oct. 1952.

Quine, W. V., "A Way to Simplify Truth Functions," *American Math. Monthly*, Vol. 62, pp. 627-631, 1955.

Surjaatmadja, J. B., *A Computer-Oriented Method for Boolean Simplification and Potential Hazard Elimination*, Fluid Power Annual Research Conference, Report No. R 73-FL-2, Fluid Power Research Center, Stillwater, Oklahoma, 1973.

Surjaatmadja, J. B., *TAB II—Program and Users Guide for the Simplification and Static Hazard Elimination of Collossal Boolean Expressions*, Fluid Power Annual Research Conference, Report No. R 74-1, Fluid Power Research Center, Stillwater, Oklahoma, 1974.

Wehrfritz, H. C., "Techniques for the Transformation of Logic Equations," *IEEE Trans. on Electronic Computers*, Vol. c-23, No. 5, pp. 477-480, May 1974.

Chapter 3

System Inputs and Outputs

3-1 INTERCONNECTION PHILOSOPHY

The purpose of a digital control system is to help the user perform repetitive tasks. The control network provides the interface between the inputs and the outputs. It alone possesses the logic of the job and the control stratagems. The design of this interface—the control network—is a factor paramount to the success of the system and machine, but it alone is helpless and unresponsive.

As in other types of control systems, a variety of elements other than the "controller" are necessary. The control network must be cognizant of the external state of the machine in which it has domain. It is essential that the control network have sufficient sensory perception to translate every condition of the mechanical system in terms of the desired future action. For example, consider the need to actuate a power cylinder when all cylinders are in a particular position. A human operator would inspect and detect this condition and would then activate the intended power cylinder. However, in order for the control network to be able to know the present condition of the mechanical device or system, "input sensors" are needed to provide this information.

The signals provided by the input sensors must be in a medium that can be utilized by the controller, i.e., liquid, gas, mechanical, or electrical. These input

SYSTEM INPUTS AND OUTPUTS

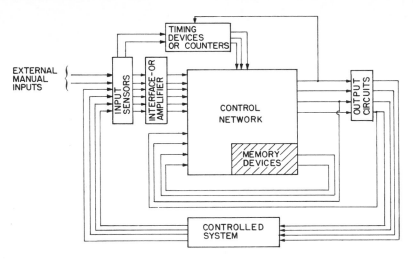

Figure 3-1 Generalized scheme of a fluid logic system.

signals provided by the sensors are often referred to as the "machine state" inputs to the system as opposed to "manual" inputs. The interfacing of the inputs with respect to the rest of the control system and particularly a fluid logic system is illustrated in the schematic shown in Figure 3-1.

The control network continually compares the current state of the mechanical system with the desired or "prescribed" condition before it gives the command for the "next" operation. To initiate this command, an "output" is activated to achieve the next operation. Like the input, the output generally requires an output circuit due to the medium interfacing and signal amplification needed to accomplish the output task.

"Primary outputs" are defined as outputs that directly control the system or machine and hence are the outputs actually desired by the machine designer. "Secondary outputs" are produced by memory elements or circuits and are used to represent the internal state of the network—storing information relative to the past and present state of the machine cycle. Note that in some cases primary outputs can be utilized as secondary outputs, and vice versa. The secondary outputs are fed back to the control network; together with the inputs they provide machine-state information for network decisions regarding the next primary and secondary outputs. This interconnection philosophy for a fluid logic control system is depicted in Figure 3-1.

3-2 INPUT SENSORS

Input sensors are used for the purpose of sensing or detecting the physical state of a machine parameter and representing such information into a form which is meaningful and usable to the control network. In many cases, the signals generated by input transducers need to be rectified, amplified, and transformed into network relatable terms to have any value whatsoever for the control network. For example, the signal from a photosensitive diode (electric eye) used for detecting the presence or translation of a physical body must be amplified and transformed into pneumatic form before an air logic network can utilize the information.

There is an unending list of input sensors capable of providing machine-state information to a digital control network. Such sensors can be classified in accordance with their sensing parameter or by their sensing medium. Examples of sensing parameters would include the following:

> position, velocity, acceleration, jerk, temperature, pressure, stress, torque, load, liquid level, flow, light, darkness, vibration, noise and gyration, and radioactivity.

An effective way of classifying input transducers is from a media standpoint (according to the type signal they produce), i.e., mechanical, electrical, or fluid. It is this classification which uniquely distinguishes the three basic forms of logic control systems.

Although mechanical input sensors have the greatest application on mechanical logic networks, they do play an important part in other types of systems. Examples of machines which are particularly good applications for both mechanical input sensors and mechanical networks include sorters, packagers, crimpers, and slot machines. Parameters such as weight, rotational speed, position, acceleration, and manual selection can be effectively detected and relayed to the control network through mechanical input elements. Some of the elements used to perform the input function include the following: cams, levers, governors, linkages, clutches, ratchets, kinematic chains, and index wheels.

There is a large variety of electrical sensors commercially available today. Mechanically actuated contacts represent one of the most common types of electrical sensors used in machine logic systems. More sophisticated devices, usually of a type closely related to a particular application, have also been developed. A glance at almost any machine periodical will reveal an endless assortment of such input devices, including electric eyes, pressure transducers, temperature sensors, and proximity switches. Such inputs are often needed regardless of the specific type of logic system being implemented or considered.

Fluid sensors are designed for hydraulic or pneumatic control service or both. Such sensors generally have specific restrictions on the pressure level at which they can operate as well as the type of fluid (from a material-fluid compatibility standpoint) with which they can be exposed. Although there are other limitations which might be given (such as temperature, contamination, and corrosion), fluid transducers do not suffer in application any more than their counterparts in other media.

Hydraulic and pneumatic devices have many similar characteristics even though more restrictions are imposed on hydraulic elements because the fluid cannot be exhausted to the atmosphere. Pneumatic devices are usually small, and when air is used as the medium, no provisions are needed to conserve the air. Small leakages

SYSTEM INPUTS AND OUTPUTS

from pneumatic systems are usually tolerable and therefore a large variety of design configurations are suitable in the development of pneumatic sensing devices.

Hydraulic input sensors are slightly different than pneumatic devices in that they require a drain or return line in order to save the costly hydraulic fluid. Leakage from high-pressure liquid systems is not only a nuisance but can be quite hazardous. Sensor elements for hydraulic service usually are larger in order to accommodate the larger passages which are needed due to the viscosity difference between liquid and gas—a factor important in parameter detection and signal transmission response.

Three-way and four-way spool valves offer an effective solution for an input sensor to a fluid network. These elements have unique provisions for a return line, which make them ideal for use as hydraulic input sensors. Also most valves of this type have relatively high pressure ratings, which allows them to be used on either pneumatic or hydraulic systems.

Three- and four-way valve-type fluid sensors are available in a variety of actuator styles, each of which has its own particular advantage or special feature for satisfying a specific need. Three-way valves with roller-lever actuators are shown in Figure 3-2 together with their circuit symbols. In Figure 3-2(a), it can be noted that an actuation of the sensor will result in the opening of the lower passage while blocking the upper passage, which results in the activation of the output Z. Interchanging the PRESSURE and TANK ports will produce the opposite effect—converting the sensor from a normally closed type to a normally open type. In Figure 3-2(b), a closed-center-type three-way valve is shown with its circuit symbol. Note that the valve in Figure 3-2(a) has an "open" center position that could influence the control network and hence must be shown in the symbol, whereas the valve in Figure 3-2(b) has a blocked center and need not be shown in its symbol unless it is physically utilized. A four-way valve-type sensor having a roller-plunger actuation is shown in Figure 3-3 with its circuit symbol—an "open center" type. Here, two separate outputs are produced, one being a complement of the other.

The actuator styles illustrated in Figures 3-2 and 3-3 are interchangeable, and still many others are available which are specially suited for novel applications; e.g., note the wire-sensor and the ball actuator shown in Figure 3-4.

Manual-type actuators are used in logic networks to produce start and stop signals, provide safety latches, and emergency control. No significant difference exists between manually and mechanically actuated valves. Usually the aesthetic aspect is reflected more in manually actuated elements—more decorative, smooth lines, cleanable, and easy actuation. Two of the common manual actuators—the push button and the toggle actuator—are illustrated in Figure 3-5. The various types of input sensors and manual input valves recognized by A.N.S.I. (American National Standards Institute) are given in Chapter 4, Table 4-1, in symbolic forms.

Another important actuator design for a three- or four-way valve is the pressure-actuated type. There are many applications for fluid logic systems where a signaling device is needed to indicate an overload condition or to monitor the pressure existing in a circuit. It is not sufficient to send a fluctuating-type signal to the control network—the signal must be conditioned and presented in the form of a

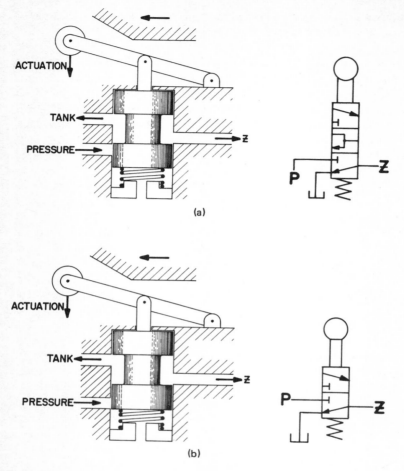

Figure 3-2 A three-way, spool-type valve, roller-lever actuator: (a) Open center; (b) closed center.

distinct pressure rise. The digital nature of the control elements of the network requires that the control signal exhibit either full pressure or tank pressure with no intermediate partial pressure values. Such requirements normally rule out conventional sequence valves. A valve design which meets all requirements for a pressure-signal input device and one which has had unqualified success in stringent applications is shown in Figure 3-6. It was developed by the Fluid Power Research Center (U.S. Patent No. 3,388,721 dated June 18, 1968) at Oklahoma State University and is known as the OSU Pressure Signal Valve. Its novel features include:

1. Is capable of monitoring the pressure in a hydraulic cylinder in order to detect pressure overload conditions.
2. Provides a "detent" action to insure a distinct "on or off" signal.
3. Offers a means of varying the pressure level at which it switches from the "off" to the "on" position.

SYSTEM INPUTS AND OUTPUTS

4 Generates a conditioned digital signal which accurately represents the pressure state of a machine power circuit.

Poppet valve-type sensors are essentially similar to the three-way spool-type sensors. The poppet units are popular, as they are generally easier to mass produce. The degree of machining accuracy of these valves is normally much less critical than the conventional spool-type valves, which minimizes the production costs. Sensors utilizing these type of valves normally have medium to high pressure ratings. Two examples of this poppet sensor type are illustrated in Figure 3-7 together with their respective circuit symbols.

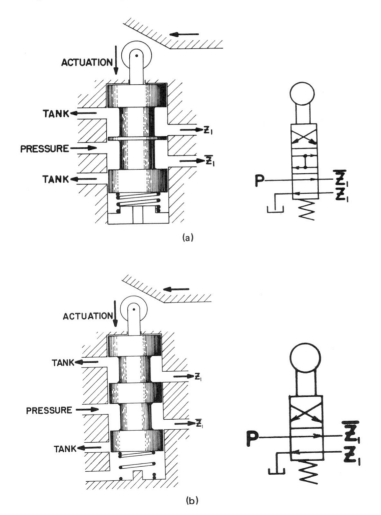

Figure 3-3 A four-way spool-type valve, roller-plunger actuator: (a) Open center; (b) closed center.

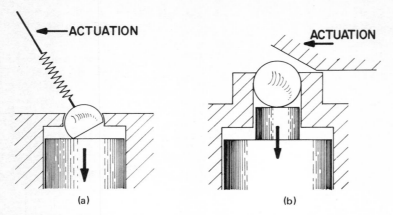

Figure 3-4 Mechanical actuators: (*a*) Wire rod actuators; (*b*) ball actuators.

Back-pressure sensors are low-pressure sensing devices which are practical when the object to be sensed is fully submerged in the media used in the control system. Because of this characteristic, back-pressure sensors are generally more practical in air logic circuits. Usually, these types of devices, which operate on pressure less than one-half of an atmosphere, are quite suitable for the detection of delicate items. In addition, since the operating pressures are low, signal amplification is required when high-pressure control networks are used. In Figure 3-8, two examples of back-pressure sensors and their A.N.S.I. symbols are shown. Note that in Figure 3-8(*a*) the presence of an object that blocks the "drain" (or sensing) port results in the activation of output signal Z. In Figure 3-8(*b*), a push button actuator serves as the "object" and the sensor has become a manually operated signal valve.

The back-pressure concept can also be employed to produce a revolutionary sensor, as illustrated in Figure 3-9. A signal X is emitted as long as the slot is not adjacent to the object port of the sensor.

The proximity sensor is simply a sophisticated back-pressure sensor. This type of sensor does not require physical contact with the object block to activate the

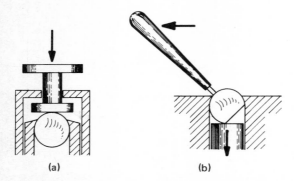

Figure 3-5 Manual actuators: (*a*) Push button actuators; (*b*) toggle actuators.

SYSTEM INPUTS AND OUTPUTS

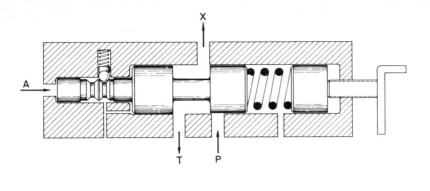

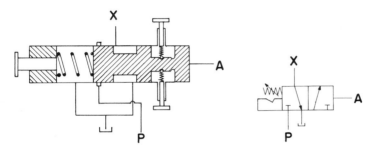

Figure 3-6 The OSU pressure signal valve.

sensor. Instead, the principle of operation is as shown in Figure 3-10. When no object is present, the flow of air is directed away from the output port, thus creating a vacuum pressure at this port. When an object passes the sensor region within a set distance (approximately 0.2 inches), a turbulent condition occurs and the airflow "feeds back" into the output port, creating an output Z equal to 1.

The interruptible jet sensor detects an object which obstructs an air jet that is focused upon an output port, as illustrated in Figure 3-11. An output Z is sensed whenever no object is present, and the output deactivates as the jet is deflected or obstructed by an object. The interruptible jet concept can be employed to produce a revolutionary sensor or even an index sensor, as illustrated in Figure 3-12. The configuration of the holes in the index wheel shown in the figure is suitable to differentiate between angles every 22½ degrees apart—other wheels can be designed to meet other requirements. Of course, the control network must keep track of the specific location of the wheel at any one time in order to know the location of the "next" desired state.

An even more sophisticated type of fluid sensor is the acoustic sensor. The principle of operation is similar to the interruptible jet sensor, except that acoustic waves are used instead of an air jet. A laminar flow "jet" is used for the detection of the acoustic waves and the output is activated whenever an object is sensed, as portrayed in Figure 3-13.

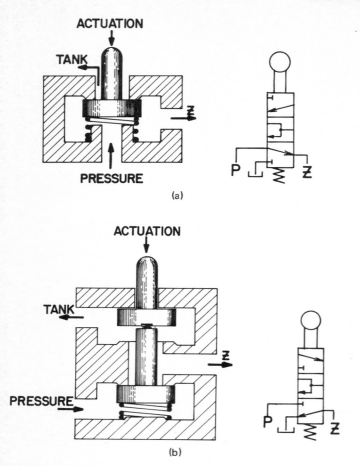

Figure 3-7 Sensors with poppet valves: (*a*) "Normally open" poppet valve sensor; (*b*) "normally closed" poppet valve sensor.

3-3 INPUT INTERFACE AND AMPLIFICATION

The selection of the input sensors is influenced by the type of control network to be employed. For example, if a high-pressure control network is being used, it is obviously desirable to select high-pressure sensors. However, as easy as it may seem, this is not always feasible. Particular situations often arise where delicate, low-pressure sensing devices must be utilized and these low-power level signals interfaced and amplified to satisfy the needs of a high-pressure control network. In this case, the interfacing method must allow amplification of the signal.

Hydraulic-to-pneumatic or electric-to-fluid interface devices are commercially available. Typical illustrations of these devices can be seen in Figure 3-14. In Figure 3-14(*a*), a fluid-to-electric interface device is shown, while in Figure 3-14(*b*) an electric-to-fluid interface (commonly known as a solenoid-operated valve) is

SYSTEM INPUTS AND OUTPUTS

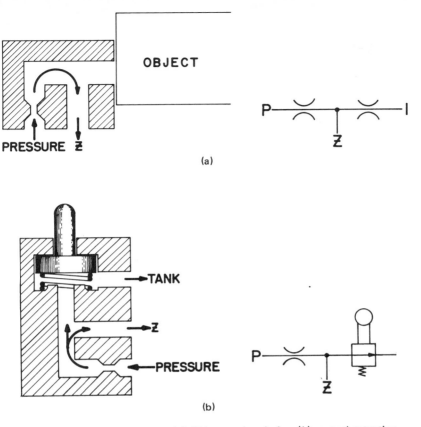

Figure 3-8 Back-pressure sensors: (*a*) Object-sensing device; (*b*) manual actuation.

illustrated. The operating features of these interfaces are quite apparent in these illustrations. Fluid-to-fluid interface and amplification devices are depicted in Figures 3-14(*c*) and (*d*). The membrane-type interface shown in Figure 3-14(*c*) is particularly useful when fluid interaction between the two media is to be avoided.

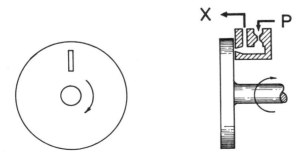

Figure 3-9 Back-pressure revolution sensor.

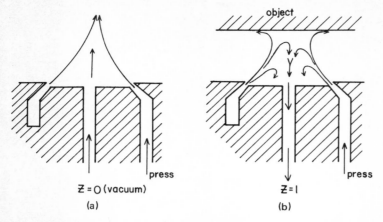

Figure 3-10 Proximity sensor: (a) No object sensed; (b) object present.

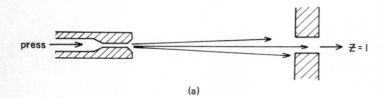

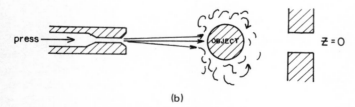

Figure 3-11 Interruptible jet sensors: (a) No object present; (b) object present.

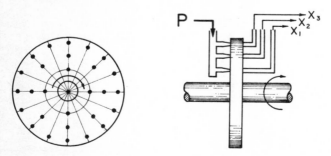

Figure 3-12 Interruptible jet index sensor.

SYSTEM INPUTS AND OUTPUTS

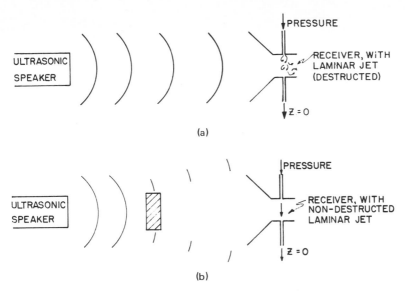

Figure 3-13 Acoustic sensor: (a) No object present; (b) object present.

The size difference between the membrane and the actuating piston of Figure 3-14(d) is used to obtain a larger pressure amplification at the interface.

3-4 TIMING DEVICES

As implied in Figure 3-1, it is sometimes necessary to delay the transmission of a signal to the control network; e.g., when "settling" or stabilizing periods are involved during which no operations should occur or when a prescribed period of time must be allocated to processes which must be performed before proceeding with the sequential operation. For such purposes, delay elements are needed and they are installed at the input of the control network or somewhere in the network itself. A particular delay circuit is illustrated in Figure 3-15(a). Note in Figure 3-15(b) that this circuit will provide a delay in the activation of the signal, but not for the deactivation of the signal. If the duration of the signal is less than the preset time delay, no delayed signal will be produced.

Mechanical timing devices are also popular. One such device is the rotating cam "absolute" timer shown in Figure 3-16. Here, the input signal a activates an external motor that rotates the cam; and, following some preset time delay, the cam actuates the fluid sensor to produce the delayed input signal b. The accuracy of the time-delay period depends upon the accuracy of the motor itself. Rotating cams can also be used to give "relative" timers—timers that record time relative to the cycle of the machine. This application is more commonly known as "sequence programmers," and one particular illustration is shown in Figure 3-17. The sequence programmer counts the number of times the input signal a is activated before actuating the fluid sensor which produces the input signal b.

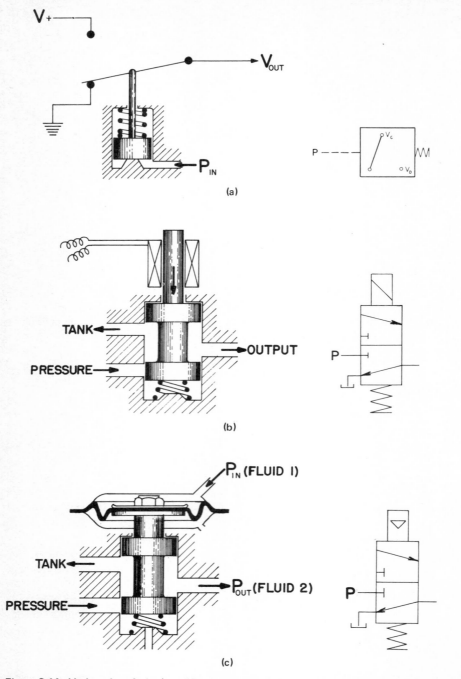

Figure 3-14 Various interfacing/amplification methods: (*a*) Fluid-to-electric; (*b*) electric-to-fluid; (*c*) fluid-to-fluid, with oversized membrane.

SYSTEM INPUTS AND OUTPUTS

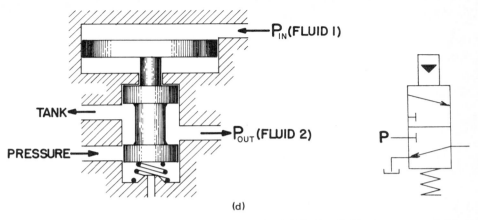

Figure 3-14 Various interfacing/amplification methods: (*d*) Fluid-to-fluid, piston actuated. (*Continued*)

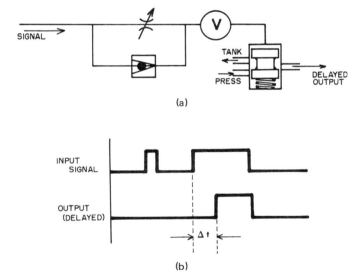

Figure 3-15 Time-delay circuit: (*a*) The circuit; (*b*) its response.

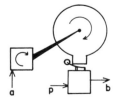

Figure 3-16 Cam timer (absolute timer).

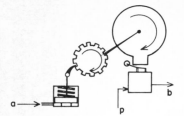

Figure 3-17 Cam timer (relative timer).

Electric or electronic timers are used when more accurate and sophisticated timing functions are required. However, when electrical timers are used, interface problems will arise for networks other than electrical.

3-5 OUTPUT CIRCUITS

Unless the desired output is a blast of gas or liquid, the output signal from a control network must be transformed into some other form—mechanical or electrical. Pressure switches are common output devices for achieving the fluid-to-electric transformation. In order to restrain the subject matter, only the fluid-to-mechanical output devices will be discussed and these are referred to as power circuits.

Basically, a power circuit provides the muscle for the control network (the brain) to accomplish its intended mission. Generally, the elements of a control network are miniaturized components and do not possess the fluid-flow capacity to provide the power-distribution function of the system. Hence, specific power circuits must be available which are capable of accepting a low-power-level signal from the control network and transforming this signal into mechanical power—linear or rotational.

There are many different versions of power circuits. Fundamentally, they consist of a power valve which can be actuated by the pressure level transmitted by the control network. In addition, the power valve with its own power source is connected to a fluid motor either linear (a cylinder) or a rotary (or oscillating) motor. By the proper application of pressure signals, the direction of motor movement can be controlled.

The power circuit shown in Figure 3-18 utilizes a spring-return cylinder and a two-position, three-way, spring-offset power valve. The cylinder will retract when signal Z is not present and will extend when signal Z exists. The circuit illustrated

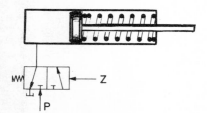

Figure 3-18 Spring-return cylinder with three-way, spring-offset valve.

SYSTEM INPUTS AND OUTPUTS

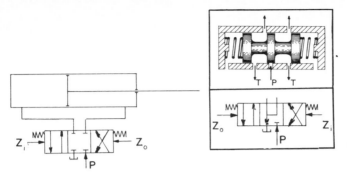

Figure 3-19 Double-acting cylinder with three-position, four-way, spring-centered valve. Insert: Open-center-type valve.

in Figure 3-19 utilizes a spring-centered, four-way, three-position valve. The center position can have almost any port-connecting configuration—two types are shown in the figure together with a cross-sectional view of a typical open-center-type valve. Figure 3-20 shows the use of a two-position, four-way, detent valve. From the cross-sectional view, an appreciation of the detent function of the valve can be obtained.

Other types of power circuits are illustrated in the next section. It should be realized that the machine, its operating environment, its design mission, and its designer all have a major part in determining the "best" power circuit configuration—these factors explain why there are different machines on the market designed to accomplish the same job.

3-6 THE POWER AND INPUT SYSTEMS

The success of a fluid logic system depends to a large extent upon how effective the input and output devices of the system are. As has already been illustrated and discussed, there is a vast assortment of such devices for the logic designer to select. For any given application, the designer who can marry the "best" combination of elements together will always maintain an edge on his competitors.

The reader should know by now that the control network is helpless without a motor/input system. It must rely upon the sensing elements to provide critical machine-state information, and it must depend upon the motor elements to carry

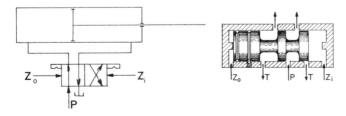

Figure 3-20 Double-acting cylinder with four-way, two-position, detented valve.

out the tasks for which the control network is programmed to coordinate and govern. Although the outputs and inputs introduced in this chapter were restricted to fluid-oriented devices, the designer must remain free to select the right elements for his application.

In order to appreciate the various motor/input systems illustrated in this section, it would help if the reader possessed a vivid imagination. He should remember that each power element would normally be connected to some mechanism which can perform some useful task—to a valve plunger, a machine table, a clamper, a furnace door, a spindle, etc. Likewise, instead of the sensors being in immediate contact with the motor element itself, as illustrated here, they may be associated more directly with the door, the table, or the spindle. Furthermore, the reader should never lose sight of the fact that electrical or mechanical sensors in conjunction with fluid-input interfacing elements have a major application value in fluid logic systems. Although a back-pressure sensor may serve very well to detect the presence of a box or machine part, an electric eye with the proper interface might be even more advantageous.

The detent power system shown in Figure 3-21 offers a simple configuration for applications such as a clamping circuit. To sense the load on a bucket or clam shell, the system illustrated in Figure 3-22 should be considered. There are many applications where multiple machine operations are performed while a part is moved in discrete steps along a linear path. A motor system such as the one shown in Figure 3-23 with its position-sensing valves might satisfy such a need.

A cam follower system portrayed in Figure 3-24 has many applications in the packaging and processing industries. With a single valve as a sensor, it can relay information to the control network relative to the rise and fall of the cam follower. One class of motor circuit which is often overlooked is the hydraulic detent system using check and pilot check valves. The versatility of control afforded by such a power system (depicted in Figure 3-25) might prove important in some critical applications.

As an example of a rotating-shaft power system, consider the output and input elements presented in Figure 3-26. Of course, to perform a useful function, the

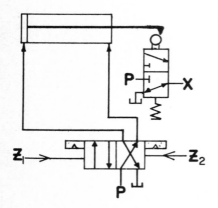

Figure 3-21 A detent power system.

SYSTEM INPUTS AND OUTPUTS

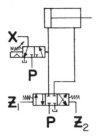

Figure 3-22 A cylinder load-detection system.

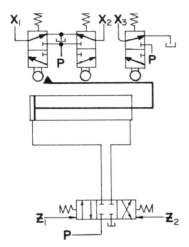

Figure 3-23 A three-point position system.

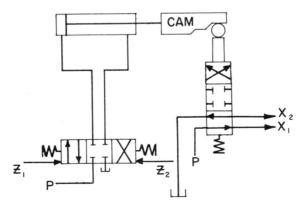

Figure 3-24 A cam-follower system.

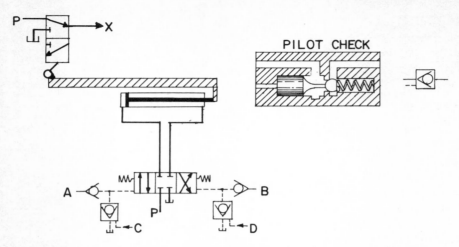

Figure 3-25 A pilot check valve system, where extend $= ADC'$; retract $= BCD'$; and hold $= CD$.

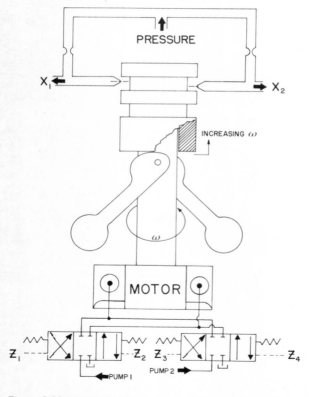

Figure 3-26 A rotational speed-governing system.

SYSTEM INPUTS AND OUTPUTS

shaft must be connected to a spindle or another type of work piece. Two separate pump systems are shown in order to illustrate the capability of the power system to vary the speed of the shaft upon command. The speed-sensing head is capable of detecting several different speed settings—a necessity in some applications.

3-7 APPLICATION NOTES

A logic system designed to satisfy a given specification must be effective, dependable, and complete. To accomplish such an important task, the designer must consider all possible situations that might jeopardize the machine's functional integrity and reflect discredit upon the logic system involved. Completeness of a logic description contributes immensely to the development of a fault-tolerant machine, even though added statements to achieve completeness may increase the complexity of the resulting control system. However, the reader is advised to be critical of gross redundancies and realize that there are applications where it is advantageous to disregard possible machine states that have a very low probability of being executed in the controlled cycle of the machine. For example, machine states that result from sudden power failures, mechanical breakdown, or other emergency situations. Unless there is actual danger to life or property, manual resetting procedures can always be incorporated and employed to correct such rare occurrences of interruptions.

The design of a logic system is highly influenced by the input and output devices that have been selected for the system. For example, a system that utilizes manual push buttons will require different considerations than a system that uses mechanical sensors as signal feedback devices from a reciprocating cylinder. In addition, some power circuits require a totally different sensor specification. For example, a power circuit that uses a spring-return cylinder does not require a return-signal sensor; on the other hand, power circuits that employ a detented four-way valve might only need a short duration signal in order to trigger a clamping operation. Another influential aspect is the type of cycle itself. For example, if it is desired that a logic cycle terminate after each complete cycle, the selection of the sensor might be completely different than for a continuously repeating cycle or one that can be changed by a selector valve.

The following points might serve as guidelines for formalizing a logic-control system specification:

1 Determine the main task of the machine or system.
2 Establish the type of logic system that will be used for the control network.
3 Design the mechanisms to which the power systems will be connected.
4 Select the appropriate input and output devices needed for the power systems.
5 Formulate the precise word statement of the problem in terms of the functioning of the selected input and output devices.
6 Note any special considerations or possible irregular situations that are to be included or avoided in the operation of the control system or machine.

In order to illustrate the formulation of a logic system specification and present the arguments and thinking needed to select input and output elements, the logic associated with a drilling operation will be considered. The main task of the proposed machine is to perform a drilling cycle which can be described as follows:

> Steel blocks of equal size are to be drilled at the geometric center to a depth of 3 inches using a ¼-inch bit. The blocks are to be conveyed to the drilling machine by means of a belt. The blocks are to be placed into position and clamped by proper jigs and fixtures automatically. After each drilling, the blocks are then ejected and placed on another conveyer belt for further processing within the plant.

Since the main task of the machine has been established by the above statement, the next item which must be considered is the selection of the type of logic system suitable for the proposed machine. In many cases, the selection is highly dependent upon the situation that exists within the plant where the machine will be installed; for example:

1. What type of control networks are common on other machines? Uniformity of control equipment may save materials and tools and avoid the necessity of employing maintenance people having varied fields of expertise.
2. What power supply is available in the plant where the machine will be installed?
3. What is the nature of products within the plant which could have an influence on the type of control system—particularly regarding product flammability and ignition temperature?
4. What is the environment which will surround the machine—take particular note of vibration, noise and temperature?
5. What restrictions must be considered regarding the delicate handling of the product involved with the machine?

In relation to the above considerations, the following factors and arguments are worth noting. If pneumatic controls were used throughout a plant, it would seem advisable that the control of the proposed drilling machine be made pneumatic also. Utilizing power and sensor elements that are familiar to current personnel helps to avoid the necessity of providing special training in a new technology and purchasing and installing new power systems. When dealing with products which are highly combustible, sensors and control devices that might produce sparks must be avoided or be shielded environmentally. Also, when the temperature surrounding the control system is critical, caution must be exercised in selecting each particular type of element. However, most control devices can operate within the temperature range of 0° to 160°F, while special devices have been designed to operate in an even broader range, e.g., $-75°$ to $300°F$. Ceramic-type fluidic devices have also been developed which can resist temperatures in the extreme burning range. The final point involved in the selection of the type of control medium is the consideration of the product itself. Delicate products may need low-pressure sensing and "kid glove"-type power

SYSTEM INPUTS AND OUTPUTS

systems, which in turn might call for low-pressure control systems and generally the need to solve interface problems.

For the drilling-control problem, it will be assumed that no delicate handling problems exist and that the use of high-pressure pneumatic systems is appropriate and compatible with the conditions in the processing plant. The next step is the design of the machine layout and the selection of the input and output circuits. This step calls for a skillful designer, as the correct selection of parts and circuitry can significantly reduce the cost of the machine itself as well as its installation and future maintenance costs.

Assume now that the drilling machine has been designed and that it exhibits a relationship of parts and circuit configuration as illustrated in Figure 3-27. In this particular design, power cylinder A is utilized for moving the steel blocks from the conveyor belt onto the drilling bench. A three-position, four-way valve with a

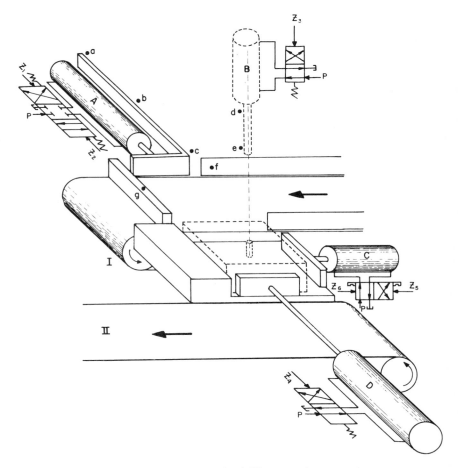

Figure 3-27 Design of power circuit for the drilling operation.

blocked center is used to hold the block in position before clamping by means of cylinder C. Cylinder B performs the drilling operation, utilizing preset pressure for the particular drilling operation. A detented four-way valve is used on cylinder C, since this cylinder is supposed to remain actuated for a relatively long period of time. A spring-return cylinder is used for B in order to ensure automatic retraction when power failure occurs. Cylinder D may be considered a normally extended power cylinder. This cylinder will remain extended unless signal D is activated. In order to produce a relatively large force, cylinder D should have a larger diameter than cylinder A. Retracting of cylinder D occurs only temporarily and longer retraction time periods can be achieved either by proper timing devices or by activation of signal D through a series of machine states.

To satisfy the requirements for input devices, assume that mechanical, normally closed sensing elements are applicable for sensing the state of the machine. The points where such elements are needed are indicated by small alphabetical letters in Figure 3-27. The problem description or machine specification can now be reformulated in relation to the machine layout of Figure 3-27 as follows:

1. Reset the system by retracting cylinders A, B, and C, which is achieved by activating outputs Z_2 and Z_6 and deactivating Z_3. Therefore, at this stage, inputs a, b, c, and d are activated; signal e is OFF; and the signals f and g are unknown or indeterminate.
2. If the sensor g is activated, which indicates that a steel block is present, activate Z_1 until signal b is OFF.
3. After a delay of 0.5 seconds, activate cylinder C by excitation of signal Z_5, and following another delay of 0.5 seconds, start the drilling by activating signal Z_3.
4. Signal e indicates that the drilling operation is almost completed (that the depth has almost reached 3 inches). A mechanical limiting device is utilized to limit the depth of the drilling from being over the intended measure; hence a sufficient delay (e.g., 1.0 seconds) may be incorporated upon signal e, which is then followed by the retraction of the drill by means of the deactivation of signal Z_3.
5. As soon as signal d is sensed, activate outputs Z_1, Z_4, and Z_6 simultaneously for transporting the block to the second conveyor. Signals Z_1 and Z_4 are held until signal c is OFF, indicating that the block has fully reached the second conveyor.
6. Still holding the output signal Z_4 in the ON position (to permit the steel block to move downstream on the conveyor), retract cylinder A by activating Z_2 until signal a is sensed. Repeat cycle by starting at Step **2**.

The above steps are the main ones needed to describe the machine's operating cycle. No alternate situations have been considered. For example, if after cylinder A has been extended sensor f is activated during a later state, it may be necessary to stop conveyor I to avoid any rubbing between the conveyor and the upstream steel blocks (sensor f indicates the presence of a steel block in the location shown). The conveyor may be started again after cylinder A has reached its fully retracted position. A more critical situation occurs when the size of the steel block is either larger or smaller than intended. By using suitable sensing and output devices, proper action can be performed to avoid almost any problem.

SYSTEM INPUTS AND OUTPUTS

At this time, it might be important to restate that the optimality of a logic system depends highly upon the selection of the input and output circuits, the design of the controlled system layout, and the design of the control network itself. Many logic network design techniques have been developed and will be presented; however, the selection of input/output elements and the layout of the machine will reflect the intuitive skills of the designer.

PROBLEMS

1. A two-cylinder system is constructed as shown in Figure P3-1(a). Cylinder A actuates cylinder B in the horizontal direction, while cylinder B is installed vertically as shown. A felt marker is installed at the actuator tip of cylinder B. The felt marker is supposed to draw the following:
   a A square, as shown in Figure P3-1(b).
   b A triangle, as shown in Figure P3-1(c).
   Determine the locations for the necessary input sensors and label them accordingly. Formulate the control network specification for each scheme.

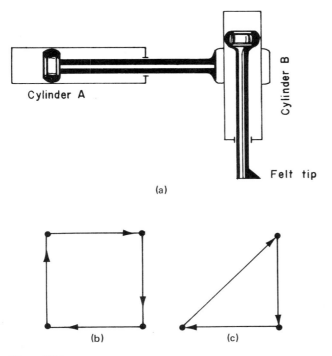

Figure P3-1

2. Determine the input locations for the two-cylinder system of Problem 1, if it were to draw the patterns shown in Figure P3-2. The numbers on the patterns indicate the sequence which must be followed. Formulate the control-network specification.

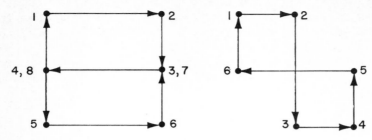

Figure P3-2

3 A conveyor belt has two sizes of boxes being transported. A box sorter at the end of the belt must recognize the box size and send them to two branch lines. Branch line *A* must receive one large box and two small boxes in that order, with branch line *B* receiving all stray or extra boxes not used to satisfy the required order for belt *A*. Design the necessary mechanism to marshal the boxes and select the appropriate hardware for the input sensors and the output system. Write the specifications for the control network in terms of alphanumeric symbols assigned to the input and output signals.

4 A proposed drill station consists of a hand-fed clamper (vice) and a drill head which can move vertically to encounter the work piece. An automatic circuit is required to clamp the part once it is loaded and both hands of the operator are on two fluid push buttons. Following the clamping of the piece, the drill advances rapidly to within ¼-inch of the work surface, then feeds slowly for two inches. After drilling, the drill head retracts rapidly and the machined part is unclamped. Sketch the general layout of the station, showing the location of the sensors and the power units selected. Prepare a complete specification for the control network in terms of alphanumeric symbols assigned to the inputs and outputs.

5 A machine table is used to clamp work pieces and move the parts into position for machining. The table is required to move in two different directions—in and out and backward and forward—in accordance with the following schedule:
   a Table moves in 2 inches and stops.
   b Then it moves forward 10 inches and stops for two minutes.
   c Then it moves in 2 more inches and returns all the way back.
   d Then the table moves in 3 more inches and the table moves forward 10 inches and stops for two minutes.
   e Then the table moves out to its original position.
   f Then the table moves back to its fully retracted position and stops for a new work piece to be clamped into place.

Draw a layout of the necessary motor and input circuits, assign alphanumeric symbols to the signals, and formulate the control network specification.

6 A shaft is supposed to rotate at a speed of approximately 100 rpm in one direction for fifty complete revolutions. Then its speed must be reduced to approximately 50 rpm for the next twenty-five complete turns, and then the shaft must be reversed in direction and go through a similar speed cycle, and then repeat continuously. Design the layout of the sensing and other units necessary for performing the task and formulate the logic specification for the control network in relation to the selected elements.

SYSTEM INPUTS AND OUTPUTS

7 A container-filling machine is required for filling a new liquid sweetener in a plastic bottle. The bottle must be fed into a weighing section, the filling tube lowered, and liquid forced into the bottle until the weight of the bottle signals the liquid to be shut off. The filler tube is then raised, the bottle is moved to a conveyor chain, and the process is repeated. Prepare a sketch of the machine and show the power and sensor elements selected. Write the control network specification in terms of the assigned alphanumeric symbols representing the input/output signals.

8 A sand-molding machine operation involves the following operations:
   a By opening a port at the bottom of a conical sand hopper, fill the mold with sand.
   b Allow one minute for the sand to settle in the mold, then close the hopper port.
   c Press the sand with a load of 5,000 lbs.
   d After thirty seconds compression time, release the press.
   e Eject the mold on a conveyor belt and position for the next molding operation.

   Design the necessary layout of the machine and select the appropriate input and output devices. Label the signals and formulate the logic specification of the control network.

9 A pallet-stacking machine is required to receive rectangular-shaped boxes (all of the same size) from a conveyor belt and stack them in one of the stacking patterns shown in Figure P3-9. Prepare sketches of the machine schematic showing the mechanisms needed to intercept, grasp, elevate, position, and release each box. Select the necessary input and output elements compatible with the recommended control network. Write the specification of the control network in terms of the symbols used to designate the required signals.

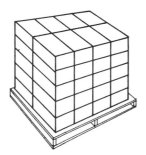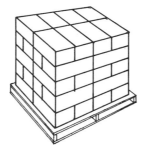

Figure P3-9

10 A pea-packer drum requires a rotating shaft drive that can be controlled in accordance with the following speed schedule:
   a Starts from rest until the shaft speed is 500 rpm.
   b The speed is decreased until the drum is rotating at 100 rpm.
   c The speed is increased to 800 rpm.
   d The speed is decreased to 100 rpm for five minutes.
   e The drum is stopped until a manual switch is actuated.

   Construct the necessary schematics of the mechanism to perform the task, select the appropriate input/output devices, and formulate the logic specification for the control network in relation to the assigned symbols for the signals.

11 A semiautomatic drill press is required to drill a hole in a steel block at one fixed location. The drilling must start two seconds after the steel block has been clamped into position. The depth of the hole should be two inches. The clamping and drilling can only be initiated by the pressing of two safety buttons, which must remain depressed throughout the operation. Releasing one of the buttons will retract the drill bit and release the clamping of the steel block. Determine the type of logic system that will be appropriate for the control network and design the layout of the mechanisms, the power circuits, and the interfaces to be used. Select the necessary input and output devices and formulate a precise word statement of the required operation which can be used to synthesize the control network.

12 Two different-size boxes are being placed on a conveyor belt in no special order. At the delivery end, the boxes must be sorted in accordance with the following schedule:
   a All small boxes to dispersal belt #1.
   b Dispersal belt #2 receives three large boxes before boxes are directed to belt #3.
   c Dispersal belt #3 receives two large boxes, then belt #2 again receives three large boxes and the process continues.
Sketch the layout of the main conveyor belt and the plan for the three dispersal belts.

   Design the necessary director mechanisms to marshal the boxes to the respective belts. Select the appropriate input/output elements, label the signals with appropriate alphanumeric symbols, and formulate the required specification for the control network.

13 A need exists for pleating three layers of paper in sandwhich fashion to form an accordion-shaped matrix for oil filters. The 12-inch-width paper rolls are fed from three separate stock rolls oriented in a vertical line one above the other. The three paper films must be fed together into the pleater section in order to form an endless accordion configuration (1-inch-deep folds). Propose a configuration for the pleater mechanism showing a pneumatic power and input sensor system. Write the control network specifications, using alphanumeric symbols to identify the various signals.

14 A speed governor is required for a shaft which will provide the following speed cycle:
   a Starting from rest, the shaft speed increases to 800 rpm.
   b The shaft speed is reduced to 200 rpm.
   c The shaft increases its speed to 500 rpm.
   d It reduces its speed to 200 rpm.
   e The engine stops.
Select the necessary input/output elements to achieve the desired speed control. Write the specification for the control network in terms of the assigned input/output signals.

15 A table for a machine tool must rotate upon its axis to a number of preselected locations where various machine operations occur. Whenever a certain position is specified by the control network, the table should always rotate in the same direction. By means of a brake and clutch system, the desired position must be

SYSTEM INPUTS AND OUTPUTS   67

anticipated and the table stopped rotating before actually reaching the position. Through a nudging process of continuously releasing and activating the clutch/brake, the required position is reached and locked into position. Lay out an effective mechanism to perform the operation and select the most appropriate input-sensing devices for the task. Choose the appropriate power-circuit elements and formulate the required control network specification.

16 A cardboard box loading station is required to handle small, finger-size, wrapped rock candy. The boxes come to the station precut, precreased, and flat while the candy is fed to the location by a gravity hopper through an overhead chute with a damper-type shutter which can be automated with a lever and linear motor. Devise the necessary mechanism which can open the boxes, fold the ends, fill the box with candy to a fixed weight, close the top, and move to the next station for banding and labeling. Select the motor and sensor elements and prepare the specification for the control network.

17 A potato weighing and packaging machine has to be automated so that it will pack 25 lbs. of potatoes in boxes. Open, empty boxes are pushed into the loading unit from a conveyor belt on to the weighing scale of the machine. When the appropriate weight is reached, the "hatch" of the gravity-fed hopper is closed gently and the box pushed on to an adjacent conveyor. Recommend the type of logic network and select the proper input, output, and power interface for the network. Formulate the necessary logic specification for the control network. Take into account irregular situations such as no potatoes, no empty boxes, emergency shutdown, and normal stop periods.

DEFINITIONS

Actuator A device for moving or controlling something indirectly.
Control medium The medium used to transmit a process and amplify signals.
Control network A network designed to analyze and formulate the operations of a machine.
Control system A system consisting of the control network, input sensors, and outputs.
Input sensor A device used to produce inputs.
Interface A device that serves as a boundary between two different signal transmission media.
Internal state An explicit description of the conditions of the memories.
Machine cycle A course of operations of the machine that recur regularly.
Machine state An explicit description of a describable stage in the operating sequence of the machine.
N-way valve A valve exhibiting N ports for directing fluid.
Primary input An input from an external source to a control network source in a medium compatible with the network.
Primary output An output that is used for the control of a machine.
Secondary input An input produced by the control network itself.
Secondary output An output used to reflect the present state of a machine.

REFERENCES

Auger, R. N., "Proximity Detection By Fluidics," *Instruments and Control Systems*, Vol. 40, pp. 108–111, June 1967.

Beeken, B. B., "Acoustic Fluidic Sensor," *Instruments and Control Systems*, Vol. 43, pp. 75–79, Feb. 1970.

Bouteille, D., and Guidot, C., *Fluid Logic Controls and Industrial Automation*. New York: John Wiley & Sons, 1973.

Budzilovich, P. N., "Fluid Logic Components," *Control Engineering*, pp. 73–80, March 1969.

Burns, P., "Timer Valve Controls Air Press Circuit," *Hydraulics and Pneumatics*, pp. 138–139, Dec. 1965.

Gesell, W. F., "Designing a Fluidic Control System," *Hydraulics and Pneumatics*, pp. 117–122, May 1966.

Gesell, W. F., "How to Use Fluidics for Industrial Control," *Machine Design*, pp. 202–209, May 26, 1966.

Fitch, E. C., "Fluid Logic," Teaching Manual Published by The School of Mechanical and Aerospace Engineering, Oklahoma State University, 1966.

Kull, R., "Versatile Air Valve Times Forging Machine," *Hydraulics and Pneumatics*, pp. 105–107, March 1966.

Kunkel, R. N., "How You Can Sense With Fluidics," *Hydraulics and Pneumatics*, pp. 94–96, Dec. 1967.

LePori, W. A., Porterfield, J. G., and Fitch, E. C., "Fluidic Control Of Seed Metering," Transactions of The ASAE, Vol. 17, No. 3, pp. 463–467, 1974.

Letham, D. L., "Transducers and Sensors," *Machine Design*, pp. 139–145, Sept. 1, 1966.

Letham, D. L., Fluidic System Design—No. 20: "Application Circuits," *Machine Design*, pp. 201–206, March 16, 1967.

Pearse, G. B., and Streeter, D. L., "Designing High Speed Machine Circuits With Fluidics," *Hydraulics and Pneumatics*, pp. 81–85, Nov. 1968.

Pinkstaff, C., Applied Fluidics—Its Impact On Industry; Part 2: "Interface Devices," *Hydraulics and Pneumatics*, pp. 114–118, April 1967.

Reason, J., "Fluidic Press Control Tests New Concepts," *Control Engineering*, pp. 74–77, April 1967.

_____, "Directional Valves For Profit-Making Designs," *Hydraulics and Pneumatics*, pp. HP1–HP32, March 1973.

_____, "Fluid Logic," *Machine Design*, Fluid Power Reference Issue, Section 12, pp. 258–266, Sept. 12, 1972.

Chapter 4

Logic Elements and Circuits

4-1 LOGIC NETWORK IMPLEMENTATION

A fluid logic system consists of appropriate input signal generators, output motor units, and a decision-making center—the control network. The implementation of the necessary inputs was discussed in Chapter 3, and attention was brought to the three basic types: manual selections, machine state conditions, and imposed time delays. The hardware items needed to implement the motor units or the power-control circuits were also covered in Chapter 3. The only part that remains is logic elements which are used to implement the control network.

The specification of the control network as described and formulated in Chapter 3 is used in the network synthesis procedure (subject to subsequent chapters) for developing the network equations (in Boolean form). These equations must be satisfied by the use of logic elements and circuitry. In order to implement the network equations with specific physical hardware, a working knowledge of available hardware as well as its inherent logic function is required. This chapter introduces the most common types of logic elements in a way that the reader can grasp their functional significance both mathematically and physically.

A Boolean equation which represents the logic requirements of a circuit is basically a nondiscipline-type function; i.e., it merely describes the logic of the circuit with no implication concerning the manner in which it can be physically employed. Thus, the expression can be satisfied by the use of electrical, electronic, mechanical, hydraulic, pneumatic, or fluidic logic elements. The only requirement for the implementation of logic network equations is that the actual hardware elements satisfy the logic functions and the interconnection be in accordance with the algebraic connectives in the equations.

Although it is realized that many different types of elements in various disciplines can satisfy the basic logic functions of the algebra, only fluid-oriented elements which are available on an industry-wide basis will be considered. Fluid logic devices other than those described will be treated as special items. A factor of utmost importance in implementing Boolean equations is that an expression can usually be satisfied by many different configurations of elements and circuit combinations. Needless to say that the specific class or type of fluid element or elements used in a given circuit would be a positive reflection of the general philosophy of the individual designer or component supplier. It is the intent of the authors that the components discussed in this chapter serve to demonstrate logic principles and not advocate specific hardware configurations.

4-2 TYPES OF LOGIC ELEMENTS

A logic element is defined as a device capable of making a zero- or one-output decision based on its input conditions. Logic elements have been devised, fabricated and marketed to satisfy:

1. The three basic functions of the algebra—AND, OR and NOT.
2. The two contractions—NAND and NOR.
3. Various auxiliary functions involving combinations—MEMORY, EXCLUSIVE OR, COINCIDENCE, INHIBITOR, AND/INHIBITOR, EXCLUSIVE OR/ COINCIDENCE, etc.
4. Supplementary logic devices—ONE-SHOT, TIME DELAY, AMPLIFIER, RELAY VALVES, etc.
5. Accessory elements such as components needed to build a complete logic circuit.

The logic elements presented and illustrated in this section are shown in several different forms:

1. A concise word description.
2. The A.N.S.I. logic symbol.
3. The N.F.P.A. fluidic symbol.
4. The A.N.S.I. fluid symbol.
5. A typical cross-section of the fluid element.
6. The truth table.
7. The Boolean equation for the element.

The A.N.S.I. (American National Standards Institute) and the N.F.P.A. (National Fluid Power Association) approved symbols represent the consensus of U.S.

LOGIC ELEMENTS AND CIRCUITS

industrial opinion and offer the most authoritative means possible for symbolizing fluid logic and fluid power components and circuitry.

AND Element (Figure 4-1)

An AND element produces an output 1-state if all of its inputs have a 1-state. For a two-input element, the truth table is:

A	B	Z
0	0	0
0	1	0
1	1	1
1	0	0

The describing equation for an AND element can be written directly from the truth table by noting the input conditions where the output requirements are satisfied. In this case, the output has a 1-state when both A and B have 1-states; thus, the equation for an AND element is simply:

$Z = AB$

A perfect AND element can be formed with a simple check valve and a pilot check as shown in Figure 4-2. An example of an imperfect AND is the pilot check valve itself as shown in Figure 4-3. The output from the pilot check valve in Figure

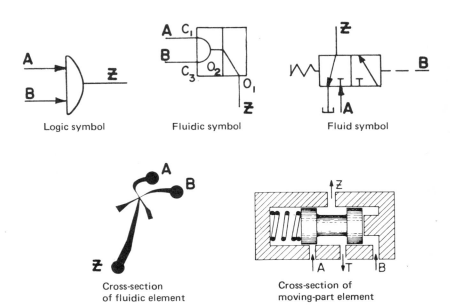

Figure 4-1 AND elements and symbols.

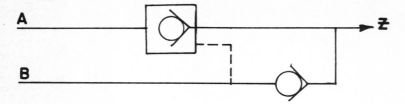

Figure 4-2 Check valve AND circuit.

4-3 is AB, but the element complement is B'; hence it is not the same as the mathematical complement $(A' + B')$ and is referred to as "imperfect."

OR Element

An OR produces an output 1-state if any or all of the inputs have a 1-state. The truth table for the OR is:

A	B	Z
0	0	0
0	1	1
1	1	1
1	0	1

The describing equation for an OR element as indicated by the truth table is:

$$Z = A'B + AB + AB'$$

which reduces to the conventional OR equation:

$$Z = A + B$$

The simple shuttle valve OR illustrated in Figure 4-4 is the simplest OR element possible and the A.N.S.I. fluid symbol tends to convey this construction. Another OR element which is simple in construction is represented in Figure 4-5.

A combination AND/OR function element can be uniquely satisfied by the check-valve circuit shown in Figure 4-6. It is an imperfect function because the fluid complement is A', whereas the algebraic complement is $A' + B'C'$.

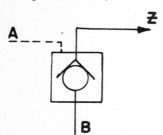

Figure 4-3 Imperfect AND circuit.

LOGIC ELEMENTS AND CIRCUITS

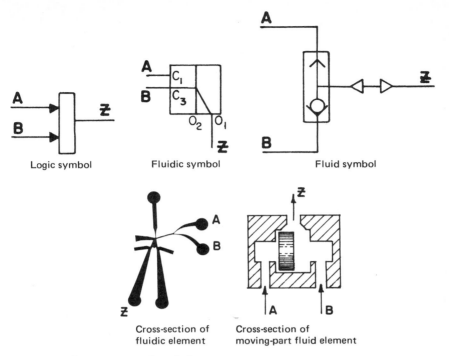

Figure 4-4 OR elements and symbols.

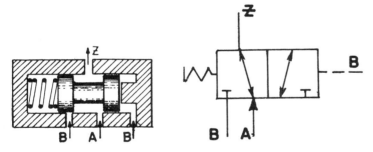

Figure 4-5 Three-way valve OR element.

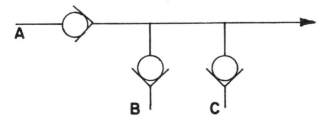

Figure 4-6 The imperfect AND/OR circuit.

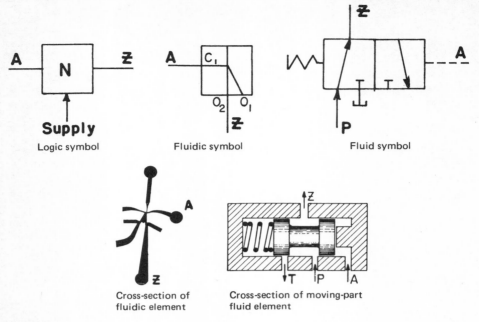

Figure 4-7 NOT elements and symbols.

NOT Element (Figure 4-7)

A NOT element produces an output 1-state only when the input has a 0-state. It is different because it is a unary element (only one input) and its truth table is:

A	Z
0	1
1	0

As reflected by the truth table, the describing equation for a NOT element is simply:

$Z = A'$

NAND Element

A NAND element produces an output 1-state if either of its inputs has a 0-state. A NAND element is a contraction of an AND and a NOT function. The truth table (two-input) for the NAND is:

A	B	Z
0	0	1
0	1	1
1	1	0
1	0	1

LOGIC ELEMENTS AND CIRCUITS

The describing equation for a two-input NAND element as given by the truth table is:

$$Z = A'B' + A'B + AB'$$

which can be simplified to:

$$Z = \overline{AB} = A' + B'$$

When dealing with moving-part NAND logic elements having a supply pressure port (as shown in Figure 4-8), it is possible to replace the supply pressure with a logic signal. Such an arrangement is represented symbolically by the circled input port as shown in Figure 4-8. This circled line should be omitted when the element

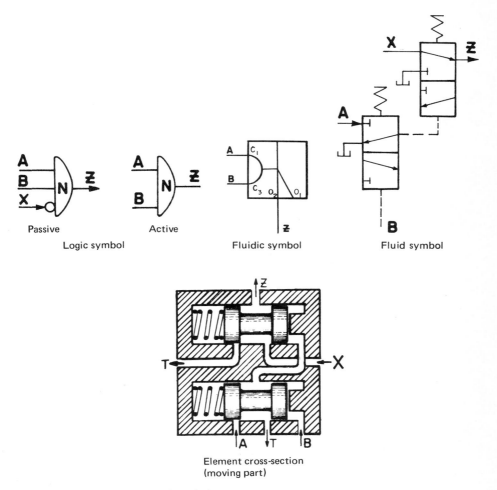

Figure 4-8 NAND element and symbols.

is utilized as an active element. The describing function of the passive NAND is:

$$Z = X \ \overline{AB} = X \ (A' + B')$$

NOR Element

A NOR element produces an output 1-state if all inputs to the element have a 0-state. Like the NAND, a NOR is also a contraction of two basic functions—an OR and a NOT. It has a two-input truth table as shown:

A	B	Z
0	0	1
0	1	0
1	1	0
1	0	0

The truth table expression for the NOR equation is simply:

$$Z = \overline{A + B} = A'B'$$

Similar to the case of NAND logic devices, moving-part NOR logic devices can also be used as passive elements. A circle is also utilized to indicate the passive

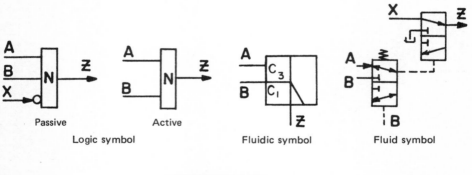

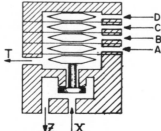

Figure 4-9 NOR element and symbols.

LOGIC ELEMENTS AND CIRCUITS

property of the particular input (see Figure 4-9). The passive NOR logic equation is:

$$Z = X \,\overline{(A + B)} = X A'B'$$

MEMORY Element

There are two basic types of memory elements on the market which are recognized by the A.N.S.I. standard (Method of Diagramming For Moving Parts Fluid Controls)—the fluid memory and the flip-flop memory.

FLUID MEMORY Element

A FLUID MEMORY has two inputs, called SET (S) and RESET (R), and one or two outputs. An output 1-state is produced by the momentary application of the SET signal and the continuous maintenance of the supply pressure. The RESET signal 1-state or the absence of the supply pressure will cause an output 0-state. Since the output of a memory element is by definition a function of the "past history," the truth table representation should include the last position or condition of the memory. The truth table for the moving-part FLUID MEMORY of Figure 4-10 with two inputs A and B, active supply pressure P, the last position of Y, indicated by Y^*, and one output Y is shown:

A	B	Y^*	P	Y
0	0	0	1	0
0	0	1	1	1
0	1	1	1	0
0	1	0	1	0
1	1	0	1	1
1	1	1	1	1
1	0	1	1	1
1	0	0	1	1
–	–	–	0	0

The truth table reveals that an output is obtained when AP exists and will remain in the 1-state as long as A and P are in the 1-state or B is in the 0-state and P is in the 1-state. The logic function can be expressed completely by the Boolean equation:

$$Y = (A + YB')P$$

As the output state is always related to its last or current state, the output Y always appears as a variable in the equation, and it is always conjoined with the complement of the reset term, (B'). Therefore, if a memory equation is to be implemented by this particular element, the terms conjoined to Y must be complemented in order to know the actual reset term. In order to demonstrate this, consider the following memory expression:

$$Y = AB'C'D + Y(B' + C' + D + E)$$

INTRODUCTION TO FLUID LOGIC

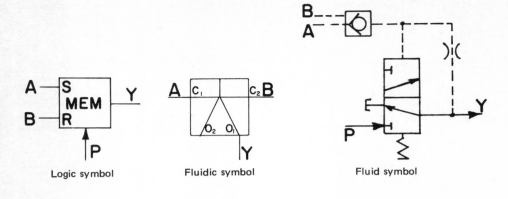

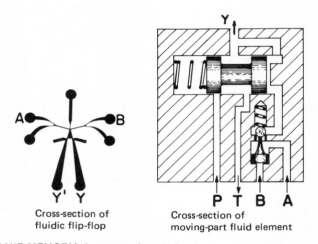

Figure 4-10 FLUID MEMORY elements and symbols.

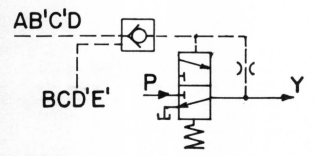

Figure 4-11 Memory implementation.

LOGIC ELEMENTS AND CIRCUITS

The classical complement of the complemented reset portion of the equation is $BCD'E'$. Using the above FLUID MEMORY element, the memory equation is completely satisfied by the application of signals shown in Figure 4-11. By definition, a fluidic-type memory device is classified as a FLUID MEMORY element because it requires the presence of a constant supply pressure to retain its memory state. In Figure 4-10, the fluidic flip-flop is shown. Note that the term "Flip-Flop" has been improperly used, and it should not be confused with the FLIP-FLOP memory element.

At this stage, it may be appropriate to mention that the characteristics of various memory elements differ from one another. For example, the truth table of the fluidic Flip-Flop is:

A	B	Y^*	P	Y	Y'
0	0	0	1	0	1
0	0	1	1	1	0
0	1	1	1	0	1
0	1	0	1	0	1
1	1	0	1	?	?
1	1	1	1	?	?
1	0	1	1	1	0
1	0	0	1	1	0
-	-	-	0	0	0

Observing the truth table, the Boolean expression which characterizes the Fluidic Flip-Flop is:

$$Y = (AB' + B'Y)\ P$$

However, because of the prevalence of the two undefined positions, the complement Y' is not represented by the algebraic complement, and is:

$$Y' = (A'B + A'Y')\ P$$

In using these types of elements, care must be taken to avoid invalid opposing signals.

FLIP-FLOP MEMORY Element

A FLIP-FLOP MEMORY has two inputs, called SET (S) and RESET (R), and one or two outputs. If the supply port is connected to a constant pressure, it is called an active element and the output of the element is transmitted by this supply energy. If the supply port is connected to a logic signal source, it is called a passive element and the output is a function of not only the memory of the flip-flop but also the logic signal. In a FLIP-FLOP MEMORY element, the memory-storing ability is not affected by the supply-pressure state or the state of the logic signal. The element has a 1-state after the momentary application of the SET signal and will remain in that state until changed to a 0-state by the application of the reset signal.

Although the state of the FLIP-FLOP is unaffected by the existence of a supply pressure or logic signal, the output of the MEMORY element certainly depends upon its supply state. If A is the SET signal, B is the RESET signal, the supply is a constant pressure P or a logic signal C, and the state of the MEMORY is Y, then the truth tables are as shown:

Constant Pressure Supply

A	B	Y*	Y	Y'
0	0	0	0	1
0	0	1	1	0
0	1	1	0	1
0	1	0	0	1
1	1	0	0	1
1	1	1	1	0
1	0	1	1	0
1	0	0	1	0

Logic Signal Supply

A	B	Y*	C=0 CY	C=0 CY'	C=1 CY	C=1 CY'
0	0	0	0	0	0	1
0	0	1	0	0	1	0
0	1	1	0	0	0	1
0	1	0	0	0	0	1
1	1	0	0	0	0	1
1	1	1	0	0	1	0
1	0	1	0	0	1	0
1	0	0	0	0	1	0

The Boolean equation which expresses the state of a FLIP-FLOP MEMORY element is given by:

$$Y = AB' + Y(A + B')$$

If a constant supply pressure exists, then the output of the element is Y. If a logic

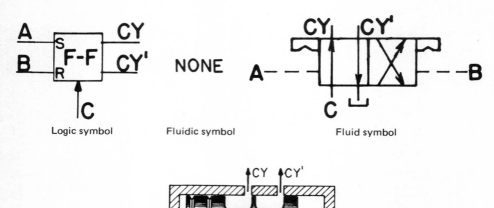

Figure 4-12 FLIP–FLOP MEMORY element and symbols.

LOGIC ELEMENTS AND CIRCUITS

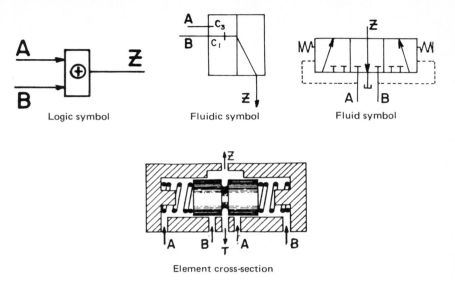

Figure 4-13 EXCLUSIVE OR element and symbols.

signal C is applied to the memory element, then the output of the FLIP-FLOP is YC.

One of the simplest FLIP-FLOP MEMORY elements available for fluid logic networks is the detent (mechanical, or magnetic, or friction) three- or four-way valve as illustrated in Figure 4-12.

EXCLUSIVE OR Element (Figure 4-13)

An EXCLUSIVE OR produces an output 1-state whenever one of the inputs has a 1-state and the other has a 0-state. Its two-input variable truth table is:

A	B	Z
0	0	0
0	1	1
1	1	0
1	0	1

The describing equation for an EXCLUSIVE OR as indicated by the truth table is:

$Z = A'B + AB'$

COINCIDENCE Element (Figure 4-14)

A COINCIDENCE element (also referred to as an EQUIVALENCE element) produces an output 1-state whenever all inputs are equal. The two-input truth table for a COINCIDENCE element is:

A	B	Z
0	0	1
0	1	0
1	1	1
1	0	0

The describing equation for a COINCIDENCE element as indicated by the truth table is:

$$Z = AB + A'B' = A \equiv B$$

It should be noted that the complement of the two-variable COINCIDENCE equation is equal to the equation of the two-variable EXCLUSIVE OR. In Figure 4-14, if the supply pressure P is replaced by a logic signal C, then the equation becomes:

$$Z = C(A \equiv B)$$

INHIBITOR Element

An INHIBITOR element produces an output 1-state when input B is in the 1-state and input A is in the 0-state. The truth table for an INHIBITOR is shown:

A	B	Z
0	0	0
0	1	1
1	1	0
1	0	0

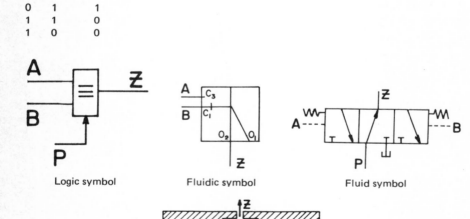

Figure 4-14 COINCIDENCE element and symbols.

LOGIC ELEMENTS AND CIRCUITS

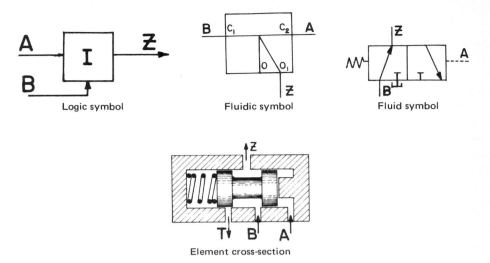

Figure 4-15 INHIBITOR element and symbols.

The describing equation for an INHIBITOR element as reflected by the truth table is:

$Z = A'B$

The INHIBITOR element differs from the NOT element in that the logic signal B is applied to the element rather than a constant pressure supply. The A.N.S.I.-approved symbols for the INHIBITOR are shown in Figure 4-15.

AND/INHIBITOR Element (Figure 4-16)

An AND/INHIBITOR element has two inputs (A and B) and two outputs (Y and Z). Output Y has a 1-state when input A has a 1-state, and input B has a 0-state while output Z has a 1-state when both inputs have 1-states. The truth table for this element is shown:

A	B	Y	Z
0	0	0	0
0	1	0	0
1	1	0	1
1	0	1	0

The describing equations for an AND/INHIBITOR element as indicated by the truth table are:

$Z = AB$
$Y = AB'$

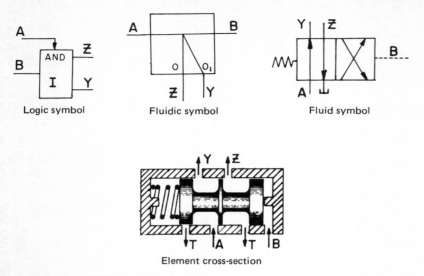

Figure 4-16 AND/INHIBITOR element and symbols.

EXCLUSIVE OR/COINCIDENCE Element (Figure 4-17)

An EXCLUSIVE OR/COINCIDENCE element has three inputs (A, B, and C) and two outputs (Y and Z). Output Y has a 1-state when input C has a 1-state and inputs A and B both have 1-states or 0-states. Output Z has a 1-state when input C and either input A or B (but not both) has a 1-state. If signal C is replaced by a constant supply pressure, it becomes an active element and satisfies the EXCLUSIVE OR/COINCIDENCE functions. The truth table for this combination element is shown:

A	B	C	Y	Z
0	0	0	0	0
0	0	1	1	0
0	1	1	0	1
0	1	0	0	0
1	1	0	0	0
1	1	1	1	0
1	0	1	0	1
1	0	0	0	0

The describing equations for an EXCLUSIVE OR/COINCIDENCE element as reflected by the truth table are:

$$Y = C(A'B' + AB)$$
$$Z = C(AB' + A'B)$$

LOGIC ELEMENTS AND CIRCUITS

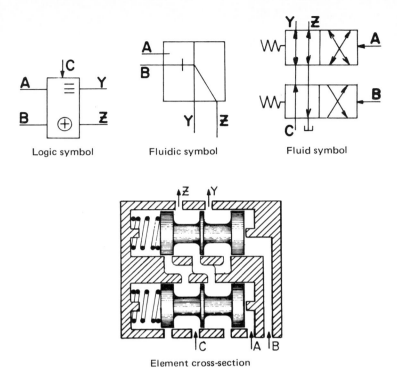

Logic symbol   Fluidic symbol   Fluid symbol

Element cross-section

Figure 4-17 EXCLUSIVE OR/COINCIDENCE element and symbols.

ONE-SHOT Element (Figure 4-18)

When the input of a ONE-SHOT element changes from a 0-state to a 1-state, the output also assumes the 1-state, but for only a predetermined period of time. Then it returns to the 0-state even though the input still has a 1-state. The time period over which the pulse exists is determined by an orifice. If an adjustable orifice is employed, the pulse length is adjustable. The truth table for this element is shown:

A	N
0	0
1	1*

*For a timed period then returns to 0.

The describing equation for the ONE-SHOT element is time-dependent but might be represented by the relation

$N = A$ (orifice dependent)

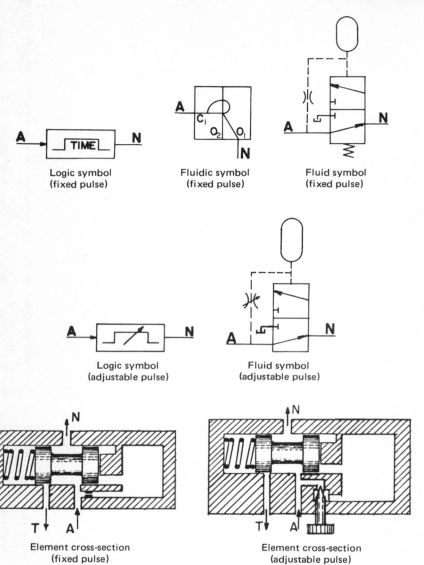

Figure 4-18 ONE-SHOT element.

which is either:

$N = A$ (⊓)   (fixed pulse)

or:

$N = A$ (⊓)   (adjustable pulse)

LOGIC ELEMENTS AND CIRCUITS

TIME DELAY Element (Figures 4-19 and 4-20)

There are two major types of time-delay elements—a TIMING IN element and a TIMING OUT element.

A TIMING IN element has a 1-state following a controlled period of time after the input assumes the 1-state. This element has a truth table as shown:

A	N^*	N
0	–	0
1	1	1
1	0	1 (delayed)

where N^* is the last condition of N. Since the output of this element is time-dependent, its describing equation must be written as follows:

$$N = A(\overrightarrow{\text{DEL}} \text{ time})$$

A TIMING-OUT element, if originally in the 0-state, changes immediately to the 1-state when the input changes to a 1-state; but when the input reverts to its 0-state, the output of the element is delayed for a controlled period of time before changing to the 0-state. The truth table of this element is as shown:

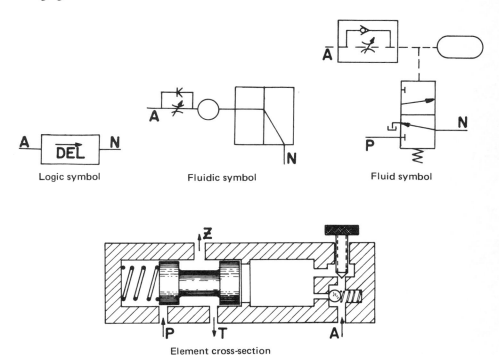

Figure 4-19   TIMING-IN delay element.

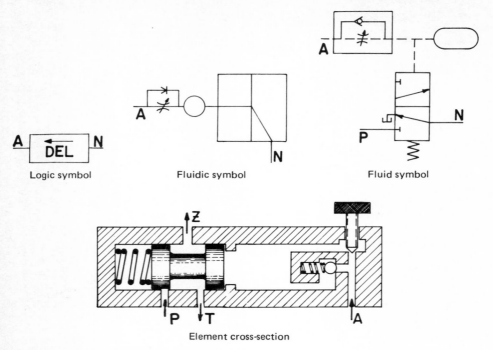

Element cross-section

Figure 4-20 TIMING-OUT delay element.

A	N*	N
0	0	0
0	1	0 (delayed)
1	–	1

Since this element is employed when a delayed output is needed in switching to the 0-state condition, the complementary expression for the element can best represent the describing equation:

$$N = A(\overleftarrow{\text{DEL}} \text{ time})$$

AMPLIFIER Element (Figures 4-21 and 4-22)

This element, consisting of one or more stages, allows an input of low energy to control an output fluid signal of a high energy level. There are two basic types—a CLOSED AMPLIFIER and an OPEN AMPLIFIER.

A CLOSED AMPLIFIER element produces a high energy level 1-state output when the input has a 1-state. The element has a truth table as shown:

A	N
0	0
1	1

LOGIC ELEMENTS AND CIRCUITS

The describing equation for a CLOSED AMPLIFIER element is simply:

$N = A$

An OPEN AMPLIFIER element produces a high energy level 1-state output when the input has a 0-state. The element has a truth table as shown:

A	N
0	1
1	0

The describing equation for the OPEN AMPLIFIER element is the complement of the CLOSED AMPLIFIER or:

$N = A'$

RELAY VALVES and Symbology

The logic elements which have already been discussed can be considered as relay valves by separating the actuating operation from the porting operation. This

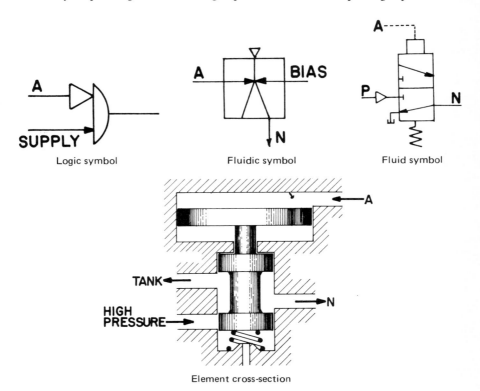

Figure 4-21 CLOSED AMPLIFIER element and symbols.

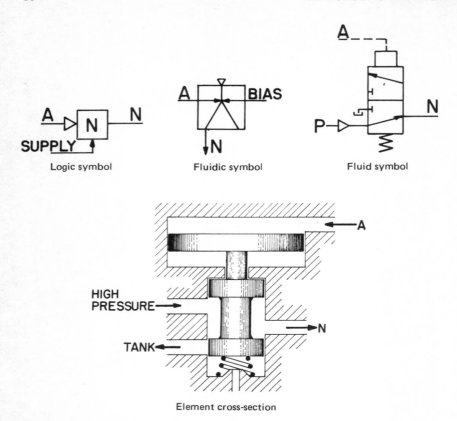

Figure 4-22 OPEN AMPLIFIER element and symbols.

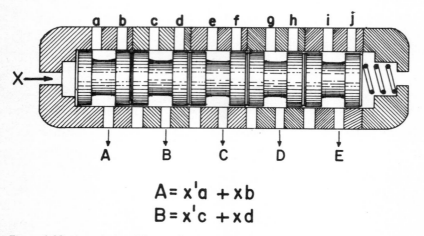

$$A = x'a + xb$$
$$B = x'c + xd$$

Figure 4-23 A typical multiport relay valve.

LOGIC ELEMENTS AND CIRCUITS

philosophy parallels that of electrical-relay technology and the conventional "ladder" diagrams. According to the A.N.S.I. Fluid Logic Standard, the ladder technique for fluid logic is called the "Detached Method of Diagramming." To utilize this method, logic elements must be represented in divided form—the actuation part and the porting part. In order to provide the reader with sufficient background to construct the A.N.S.I. "detached diagrams," the approved relay symbols for the various elements are given. Note that the "control point" for a valve is the actuation port of the valve and the "controlled flow passage" refers to the normal open or closed condition of the flow path between the outlet and inlet ports. A typical multiport relay valve is shown in Figure 4-23. It may be noted that each section of the multiport relay valve is a three-way valve. Each of its outputs can be represented as a function of all its inputs. For example, for output A its functional representation is:

$$A = x'a + xb$$

This function shows that the relay valve is a universal logic element—it can satisfy all basic logic functions, i.e., AND, OR, and NOT, each of which are achieved by connecting a to tank ($= 0$), connecting x to b, and connecting a to pressure ($= 1$) while connecting b to tank, respectively. These utilizations have been demonstrated earlier in this chapter.

CONTROL POINT Symbols

The A.N.S.I. Standard defines a complete array of symbols to depict various functions associated with relay logic operations. The following list contains most of the common symbols used in industry:

Spring-return actuator		----(RV3)
Pressure switch		----(PS 1)
Double-detented actuator	(left side)	----[RV 4A]
	(right side)	----[RV 4B]
Spring-return actuator (double control)	(opposite spring)	----(RV 5A)
	(on spring side)	----(RV 5B)

Free-floating actuator (double control, connected to a jet sensor)	(left side)	----▽ JRV6 A
	(right or bias pressure side)	----▽ JVR6 B
Memory element	(set side)	---- RV7 A
	(reset side)	---- RV7 B
Time delay (single control)		---(TR 1)
Time delay (double control detented)	(actuate)	---- TR 2A
	(reset)	---- TR 2B
One-shot (single control, automatic reset)		---(OSR 1)
Off-return relay valve	(set side)	---- RV 2A
	(reset side)	---- RV 2B
Three-position, spring-return relay valve (double control)		---◇ RV 2A
Electric-to-air (single solenoid, spring return)		(ERV 1)
Electric-to-air (double solenoid, detented)	(actuate)	ERV 2A
	(return)	ERV 2B
Electric-to-air (air AND electric signal)		▷ ERV 3

LOGIC ELEMENTS AND CIRCUITS

Electric-to-air
    (air OR electric signal)

Note:
    RV = Relay Valve
    PS = Pressure Switch

Control Flow Passage Symbols

Relay passage–closed   (exhaust is assumed)

Relay passage–open   (exhaust is assumed)

Relay passage–normally closed but open at start
    (exhaust is assumed)

Relay passage–normally open but closed at start
    (exhaust is assumed)

Relay passage–open   (nonexhausting function)

Relay passage–closed   (nonexhausting function)

Time-delay passage–closed   (exhaust is assumed, delay in opening)

Time delay passage–open   (exhaust is assumed, delay in closing)

Time delay passage–Closed   (exhaust is assumed, opens immediately, closes after delay)

Time delay passage–Open   (exhaust is assumed, closes immediately, opens after delay)

One-shot passage–closed   (exhaust is assumed, opens immediately, closes after non-adjustable time)

One-shot passage–open   (exhaust is assumed, closes immediately, opens after non-adjustable time)

One-shot passage—closed (exhaust is assumed, opens immediately, closes after adjustable length of time)

One-shot passage—open (exhaust is assumed, closes immediately, opens after adjustable length of time)

Accessory Elements for Fluid Logic Networks

Test point—identifies location

Pressure indicator—presence of pressure

Pressure gauge—provides exact pressure

Pressure regulator (adjustable, relieving type)

Filter—for particulate matter

Lubricators—for pneumatic systems

Air dryers—for pneumatic systems

Fixed resistance

Adjustable resistance

Simple check valve

Pilot check valve

LOGIC ELEMENTS AND CIRCUITS 95

Shuttle valve

Visual indicator (spring-return type, single control point)

Visual indicator (two-position detented, double control point)

General logic symbol—for functions not elsewhere specified—
  (must be labeled adequately to identify the function performed)

4-3 IMPLEMENTATION CONSIDERATIONS

The utilization of fluid logic elements for satisfying the requirements of network equations involves a number of factors which can easily jeopardize the success of a system. These factors relate to the type of elements selected, restrictions on the system response, and special machine applications. Because of the importance of these considerations, each aspect involved will be appropriately discussed.

Passive Versus Active

An active device requires constant application of a fluid supply source for its functioning, whereas a passive device assumes an active role only when the imposed signals to the device are active. A fluidic element is basically an active element because it usually depends upon the interaction of flow streams to perform logic functions. There are, of course, exceptions to this statement such as the fluidic AND and the EXCLUSIVE OR elements.

Active devices which depend upon the presence of a pressure source and particularly fluid flow to achieve their performance create problems worth noting. A pressure source at the inlet port of a moving-parts logic device poses a constant threat of hazards by "exhausting" the entire power supply. Furthermore, in resistive and fluidic-type devices where flow must persist at all times to create the logic function, the problem of energy conservation can plague the designer and the user. The use of passive devices in such systems is hence an important approach toward the conservation of energy.

In moving-parts logic, most active devices (usually devices with medium to high pressure ratings) can be used as passive devices. This is achieved by utilizing the pressure port as an input port for performing an additional AND logic operation, thus saving an AND element. For example, with the FLIP-FLOP MEMORY, a logic signal at the pressure port results in a logic AND-ing with both memory conditions. A NOT element, when utilized as a passive element, becomes an INHIBITOR, and so forth. It should be realized that due to impedance-matching problems, active fluidic devices can not be used as passive devices. This is attributed to the small, energy-conserving, resistive paths that are used at the inputs and power-supply ports of a fluidic element. Furthermore, the reader should note that passive fluidic devices also have restrictions—a general rule is not to connect two or more passive devices in series.

Fan-in and Fan-out

The term "fan-in" refers to the number of permissible inputs to a logic element. Most elements used for implementing control network equations have finite limitations on the number of input signals which they can process. The fan-in limit for some type of devices represents a real restriction which the designer is required to observe; whereas for others it depends strictly upon element size and interconnection conveniences. The need to increase the fan-in of logic elements is important, because using elements having a higher fan-in number not only reduces the number of logic elements needed in the network but also significantly reduces the switching time by reducing the number of stages (elements in series) required to satisfy a given output equation. Note that the number of stages that is required for implementing a sum or product of literals is represented by the nearest higher integer of:

$$N = \log_a M \qquad a^N = M$$

where a is the fan-in number and M is the number of literals within the sum or product. For example, for a system using three-input logic elements, a seven-literal sum or product can be implemented by two stages ($N = \log_3 7 = 1.7712$), while using two-input logic elements, the number of stages required is three ($N = 2.8073$). In general, a network equation containing complemented and uncomplemented literals can be constructed by no more than three stages, when unlimited fan-in elements are used. These three stages are:

1. The complementation stage.
2. The disjunction stage.
3. The conjunction stage.

"Fan-out" represents the number of logic elements of like kind that can be operated or controlled in parallel by a given element when all elements operate at the same pressure. Generally, there is a maximum amount of power which can be drawn by an element to operate subsequent control stages. Consequently, if the elements depend on fluid flow to perform the logic operation, the number of elements which a device can drive by its output is limited. This "fan-out" value is usually specified by the manufacturer of the logic element. Respect must be paid to the "fan-out" value because the overloading of a unit can cause serious deterioration

LOGIC ELEMENTS AND CIRCUITS

in the signal value and in switching performance. Of course, the problem of "fan-out" for medium to high pressure moving-parts logic is of little value, since the power to actuate represents an almost insignificant part of the total pressure level of the system.

Prepared Path

The response time of a fluid logic system is dependent upon the actuation time of an element and the number of stages of such elements which must be traversed before an output results. It may be difficult to reduce the actuation time of the individual elements involved, but the design of the signal path might be improved to reduce the normal response time of the output.

Besides the reduction of long signal lines, or element stages, the most obvious way to reduce signal transmission delays is to eliminate unnecessary gating. Whenever a signal must combine with another signal to produce an output, a delay can result. Such a delay is unnecessary when a gating signal is always the last signal to be energized. If a logic valve can be shifted or actuated prior to the last signal's arrival and held into "open" port position (even by the use of detent valves), there would be no delay in the output signal transmission and a faster circuit would result.

Recognizing the value of eliminating this actuator-type transmission delay in a logic circuit is credited to J. H. Cole, who conceived the term "prepared path." In his doctoral thesis, Cole found that routing an input signal directly through a "prepared path" to start the next event produces the fastest possible time for a fluid logic system. The application of the "prepared path" concept has great value in implementing control networks with fluid logic hardware, and a means of achieving this open path is worth considering.

Unless the actuation port of a logic symbol is properly identified and respected when the assignment of signals is made, it is difficult to consider the "prepared path" during the implementation process. Hence, the authors suggest that the flow passage port on the logic symbol be identified by an input arrow. Thus signal priorities can be satisfied during the initial logic diagramming stage. Since this means of signal assignment is convenient, the designer can learn to recognize the proper signals to apply to an element even during the network synthesis period. In fact, the actual form or arrangement of the network equations may be constructed to establish a methodology unique to a given designer.

4-4 EQUATION IMPLEMENTATION CIRCUITS

Once Boolean equations have been derived to satisfy the requirements of a fluid control network, they must be implemented appropriately with logic elements to obtain the actual network. The physical interpretation of logic equations requires practice in recognizing the forms of the basic functions contained in a circuit equation. Therefore, it is important to learn to associate basic logic functions with specific hardware available to the designer.

Network equations can be expressed in several different forms as follows:

1 Mixed logic forms:
   a Disjunctive—sum of products.
   b Conjunctive—product of sums.
   c Mongrel—mixture of disjunctive and conjunctive.
2 Contraction forms:
   a NOR logic.
   b NAND logic.

Of course, it is not essential that the equations be in any particular form in order to implement with hardware. Sometimes it helps to classify the equation in order to identify specific elements in the expression. Regardless of the method employed, the designer must learn to detect the "kinds of elements" needed to satisfy an equation. For example, consider the equation:

$$Z = A'B' + AD' + AB \qquad (4\text{-}1)$$

It should be recognized that the combination of terms $AB + A'B'$ represents a coincidence function, while AD' indicates inhibitions, which allows the entire equation to be implemented by the circuit shown in Figure 4-24. Consider the equation shown below:

$$Z = AB' + Z(A + B') \qquad (4\text{-}2)$$

It should be recognized that this is a MEMORY equation, because the output Z is also a variable in the equation. With a MEMORY valve, it can be implemented as shown in Figure 4-25. However, suppose that for some reason the designer did not recognize that the equation could be implemented by such a simple means and elected to use a mixed logic treatment; he would have the circuit shown in Figure 4-26(a). Another means for achieving a MEMORY function without the use of MEMORY elements is by the dual-NOR construction shown in Figure 4-26(b).

The transformation of a disjunctive form relation into NOR function form can be demonstrated with the equation:

$$Z = A'B + AB'C \qquad (4\text{-}3)$$

Complementing each term twice and the entire equation twice can be illustrated as

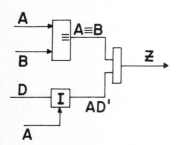

Figure 4-24 Circuit with COINCIDENCE and INHIBITOR elements.

LOGIC ELEMENTS AND CIRCUITS

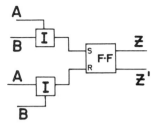

Figure 4-25 MEMORY implementation.

follows:

$$Z = \overline{\overline{\overline{A'B} + \overline{AB'C}}} \tag{4-4}$$

Performing the first complement gives the NOR form:

$$Z = \overline{\overline{(A + B') + (A' + B + C')}} \tag{4-5}$$

This equation can be implemented by the NOR logic element configuration shown in Figure 4-27. The mixed logic configuration would be as shown in Figure 4-28. The NOR logic equivalent expression for Equation (4-2) is

$$Z = \overline{\overline{(A' + B) + (Z' + A') + (Z' + B)}} \tag{4-6}$$

and the NOR-logic equivalent circuit for the MEMORY valve implementation is shown in Figure 4-29.

To demonstrate the transformation of a disjunctive form relation into NAND function form, consider Equation (4-3). The transformation is achieved by double complementing the relation and then performing the first complement (partially).

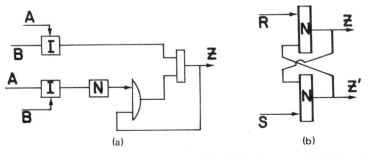

Figure 4-26 Implementation without MEMORY elements: (a) Mixed logic implementation; (b) NOR logic memory.

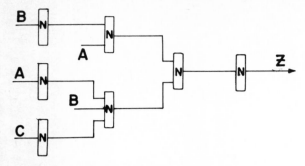

Figure 4-27 NOR logic implementation for Equation (4-3).

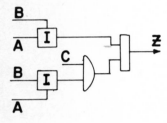

Figure 4-28 Mixed logic configuration for Equation (4-3).

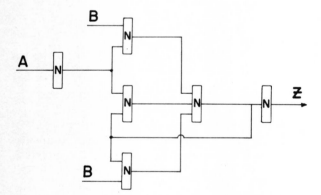

Figure 4-29 NOR logic implementation of the MEMORY Equation (4-2).

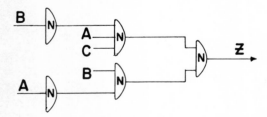

Figure 4-30 NAND form of Equation (4-3).

LOGIC ELEMENTS AND CIRCUITS

Figure 4-31 NOR logic implementation.

The operation is demonstrated as follows:

$$Z = \overline{\overline{A'B} + \overline{AB'C}}$$
$$Z = (\overline{\overline{A'B}}) \cdot (\overline{\overline{AB'C}}) \tag{4-7}$$

Equation (4-7) is the NAND form of Equation (4-3) and its implementation in NAND logic elements is as shown in Figure 4-30. To demonstrate the importance of recognizing specific elements contained in an equation, consider the following equation:

$$Z = A'B'C \tag{4-8}$$

The term $(A'B')$ should be recognized as a NOR, which would allow the equation to be written as:

$$Z = C\,(\overline{A + B}) \tag{4-9}$$

Employing the NOR element, Equation (4-9) can be implemented with only one element, as shown in Figure 4-31; but implementing with INHIBITOR logic elements would require two elements, as illustrated in Figure 4-32.

4-5 FLUID CIRCUITS

In accordance with the A.N.S.I. Standard entitled "Method of Diagramming For Moving Parts Fluid Controls," two distinct methods are recommended for presenting fluid logic diagrams—the attached method and the detached method. The attached diagram is one in which all functions and connections to component symbols are shown in the symbols. The detached diagram, on the other hand, is one in which the various functions and connections of component symbols are shown by separate symbols located in various places on the diagram.

The Attached Method of Diagramming

In the attached method, graphic symbols reflecting the expressed logic functions are used to represent the various elements. These symbols are the same logic symbols

Figure 4-32 INHIBITOR logic implementation.

presented in Section 4-2 with the valve cross-section and their corresponding fluid symbols. The attached diagram is arranged such that the symbol in the diagram facilitates the use of direct and straight interconnecting lines. The Standard recommends that the lines between the symbol inputs and outputs should be drawn horizontal or vertical with a minimum of line-crossing and with spacing to avoid crossing. The circuit itself should be arranged in functional sequence—left to right and top to bottom. This rule should be followed rigidly, if possible, as excessive line-crossing impairs the clarity of the diagram. All of the inputs to the control circuit should be shown on the left of the diagram and all outputs on the right. As there are various input sensors available today, many symbols are introduced in order to identify them. Some of these symbols are shown in Table 4-1.

To promote clarity in the attached diagram, the operational diagram (presenting the sequence of operations) can be combined and shown connected to the logic diagram. Solid lines are used to represent interconnecting flow paths between logic devices within the control circuit. Use alphanumeric designations for inputs, outputs, flow paths, and logic devices to facilitate the checking of circuits. Assign all inputs and outputs of the control circuit letters and combinations of letters and numbers (A, B, C, X, Y, Z, etc., or $A1$, $A2$, $A3$, $Z7$, $Z10$, etc.). To further identify the inputs from the outputs, it is recommended that the inputs be assigned a letter from the first half of the alphabet with the outputs assigned a letter from the last half of the alphabet (0 and 1 are omitted). To identify the flow path, A.N.S.I. suggests the use of a detailed logic diagram to show each interconnecting flow path (between inputs and outputs and logic devices and between logic devices) and number consecutively. Identify each logic device with a numerical designation, starting with the number 1 and numbering consecutively. Explanatory notes should be added where necessary to clarify the functional description. The sequence of operation should be listed in the order in which they occur with reference to input conditions and resulting output conditions.

The Detached Method of Diagramming

This method utilizes relay valve notation for identifying the various elements. The symbols for this diagramming method were presented in Section 4-2. Essentially, the method presents a (logic) valve in a separated form—it separates the actuating portion of the valve from its passage portion.

Most valves that are used in performing logic functions have two practical passages—the one that is open (or passing) and the one that is closed (not passing or tanked). An open passage would close whenever its related control port is actuated. On the other hand, a closed passage will open if its related control port is pressurized. For example, consider the detached (ladder) diagram shown in Figure 4-33.

In line k, the spring-return actuator is labeled RV5. The passage shown in line l is a closed passage and is also labeled RV5, indicating that this is a passage of a valve that is actuated by the actuator RV5. Note that the actuation of RV5 would result in the opening of the passage. An l is inserted in parentheses in line k to indicate the location of the particular passage. In line m, RV5 is also represented by

LOGIC ELEMENTS AND CIRCUITS

Table 4-1 A.N.S.I. Symbols for Input, Output, and Peripheral Equipment

ACTION	SYMBOLS	ACTION	SYMBOLS
GENERAL INPUT SYMBOLS		**MANUAL INPUTS**	
Terminals, Implied Exhausting Function		3-Way, Normally Open, Spring Return, Palm Button	
Terminals, No Implied Exhausting Function		3-Way, Detented, Open at Start, Toggle	
Operator & Bridge: – Flush Extended, Half Guarded or Guarded		2-Way, Normally Closed, Treadle Foot Operator. Spring Return.	
– Palm Button		4-Way, 5-Port (with Exhaust), Spring Return, with Alternate Output Exhausted	
– Toggle: Solid Arrow Indicate Start Position		**MECHANICAL INPUTS & SENSORS**	
Spring Return Action		Limit Valve, Normally Closed, Not Actuated at Start of Cycle	
Detented (by dotted line)			
Treadle Foot Operator		Limit Valve Normally Open, Not Actuated at Start of Cycle	
Mechanical Operator			
Float Operator		Pressure Sensor, Normally Closed, Actuated at Start of Cycle	
Air Jet			
Pressure Sensing Operator		Limit Valve, Detent, Closed at Start of Cycle (two way, no exhausting function)	
Temperature Sensing Operator			
Flow Sensing Operator		Four Way, 4-Port Float Actuated Valve. Actuate on Rising Level.	

the open passage shown. The activation of RV5 will cause the closing of the particular passage, and this is indicated by the underlined entry in line k.

Relay valves Nos. 6 and 10 are detented relay valves. Such valves have two opposing actuators, which in the ladder-diagramming procedure are represented by A and B denotations. Observing line k, it is noted that an open passage related to the A control port of RV10 exists. As the actuation of the A-port will result in the closing of the passage, the k entry of line m is underlined. Furthermore, as the B-port opposes the operation of the A-port and thus causes the particular passage to open, k is also represented in line n (without underline). The bracketed entries are

Table 4-1 A.N.S.I. Symbols for Input, Output, and Peripheral Equipment (*Continued*)

ACTION	SYMBOLS	ACTION	SYMBOLS
POWER VALVE CONTROLS POINTS		**VALVE ACTUATOR SYMBOLS**	
Spring Return, Two Position, Hydraulic Valve No. 1	– – (H1)	Spring	
Double Pilot, Two Position, Detent, Pneumatic Valve No. 2, Port A	– – (P/2A)	Manual	
		Lever	
Double Pilot, Three Position, Spring Centered, Hydraulic Valve No. 3, Port B	– – (H/3B)	Push Button	
		Treadle/Pedal	
Double Control Point, Two Position, Spring Offset Pneumatic Valve No. 4, Port A, Opposite the Spring	– – (P/4A)	Mechanical	
		Detent	
Port B, Spring End	– – (P/4B)	Reversing Motor	(M)
Pressurized at Start of Cycle	– – (P/5)	Solenoid	
Check, Simple		Pilot Pressure	
Check, Pilot Operated to Open		Actuation by Release of Pressure (pneumatic)	
Two-Way, Infinite Position Valve		Pilot (pneumatic) OR Solenoid	
Two-Way, Two Position Valve		Pilot (hydraulic) AND Solenoid	
Three-Way, Two Position Valve			
Four-Way, Three Position Valve			

only used when an opposing actuator exists and merely indicates where such an opposing actuator is located in the diagram.

An OR-ing function is most conveniently achieved by the use of a shuttle valve due to the simplicity of its construction. However, this structural simplicity cannot be reflected on the diagram when the previous port-passage concept is used. By using this concept, the representation of the shuttle valve is as shown in Figure 4-34(*a*). The representation of its three-way valve counterpart is less complex and is shown in Figure 4-34(*b*). In order to avoid this unnecessary addition in complexity, A.N.S.I. provides a special symbol for the representation of the shuttle valve, which can be observed in Figure 4-34(*c*).

LOGIC ELEMENTS AND CIRCUITS

Table 4-1 A.N.S.I. Symbols for Input, Output, and Peripheral Equipment (*Continued*)

ACTION	SYMBOLS	ACTION	SYMBOLS
PHERIPHERAL EQUIPMENT			
Pump, Uni-Directional (hydraulic)		Accumulator, Spring Loaded	
Pump, Bi-Directional (hydraulic)		Filter	
Pump, Variable Displacement (hydraulic)		Separator, Normal Drain	
Motor, Uni-Directional (pneumatic)		Filter-Separator, Automatic Drain	
		Lubricator	
Pump-Motor, Variable Displacement, Hydraulic		Heater	
Oscillator, Hydraulic		Cooler	
Electric Motor		Pressure Indicator	
Accumulator		Temperature Indicator	
		Flow Rate Meter	

Illustration of Fluid Diagrams

In order to convey the technique which has been developed and approved by A.N.S.I. for diagramming fluid logic circuits, the drilling-machine problem used in Chapter 3 for demonstrating the writing of network specifications will be diagrammed in both the attached and detached formats. The description of the problem has been reformulated as follows:

1 Start the system by automatically retracting cylinders A, B, C, which is achieved by activating outputs Z_2, Z_4, and Z_6.

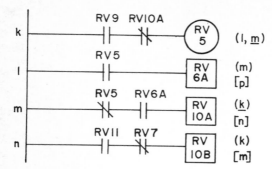

Figure 4-33 A ladder diagram.

2 If the sensor g is activated, which indicates that a steel block is present, activate Z_1 until signal b is OFF.
3 After a delay of 0.5 seconds, activate cylinder C by the excitation of signal Z_5, and following another delay of 0.5 seconds, start the drilling operation by activating output Z_3. At this time, the motor of the drill must be turned ON by activation of the pressure switch *PS1*.
4 Signal e indicates that the drilling operation is almost complete. A mechanical limiting device is utilized to prevent the depth of the drilling from being over the intended measure; hence, a sufficient delay (e.g., 1.0 seconds) may be incorporated upon signal e, which is then followed by the retraction of the drill by means of signal Z_4.
5 As soon as signal d is sensed, outputs Z_1, Z_6, and Z_7 are activated simultaneously for transporting the block to the second conveyor. Signals Z_7 and Z_1 are

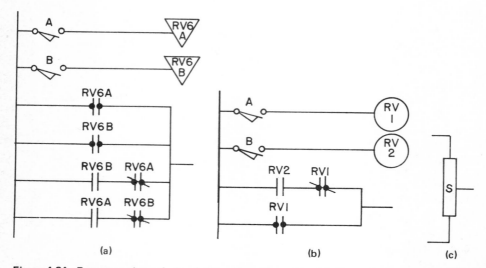

Figure 4-34 Representation of the OR-function $(A + B)$: (a) Shuttle valve representation; (b) three-way valve representation; (c) A.N.S.I. symbol.

LOGIC ELEMENTS AND CIRCUITS

held until signal c is OFF, indicating that the block has fully reached the second conveyor. At this instance the drill motor is stopped.

6 While the output signal Z_7 is still in the ON position, cylinder A is retracted by activating Z_2 until signal a is sensed. The cycle is repeated by starting at Step 2.
7 If the cylinder A has been extended and $LV6$ (Limit Valve 6, producing signal f) is activated, stop conveyor I until cylinder A is fully retracted (a is ON).
8 The STOP signal will merely stop the conveyor belt I, and hence the sequence will automatically terminate at the end of the cycle. The necessary input-output circuits are shown in Figure 4-35. Note that the two push buttons (the START-STOP button) are not shown.

In this figure, the ports of the power valves are designated by $P1A$, $P1B$, and so forth, where A and B refer to left and right, respectively. The limit valves used ($LV1$, $LV2$, etc.) are valves that produce one "true" signal, namely, a, b, c, etc. By using one of the synthesis techniques presented later in this book, the following circuit equations can be derived:

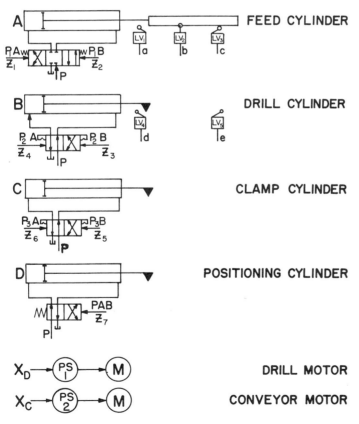

Figure 4-35 Output circuits for the drilling operation.

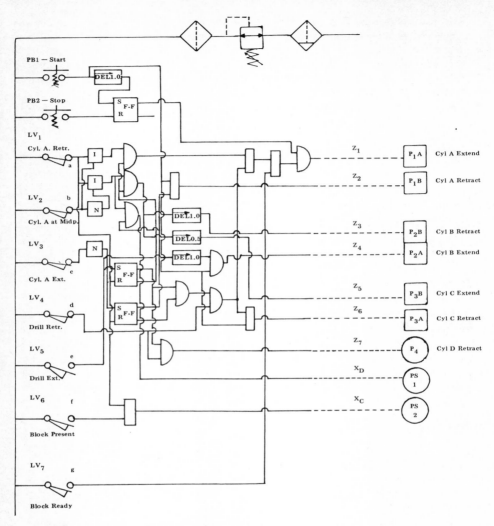

Figure 4-36 A.N.S.I. attached logic diagram.

$$Z_1 = (g + a'bY_2' + dY_2Y_3') \cdot Y_1$$
$$Z_2 = \text{START} + Y_3$$
$$Z_3 = a'b'Y_2' \ (\overrightarrow{\text{DEL}}\ 1.0\ \text{sec.})$$
$$Z_4 = e\ (\overrightarrow{\text{DEL}}\ 1.0\ \text{sec.}) + \text{START}$$
$$Z_5 = a'b'Y_2' \ (\overrightarrow{\text{DEL}}\ 0.5\ \text{sec.})$$
$$Z_6 = \text{START} + dY_2Y_3'$$
$$Z_7 = dY_2$$
$$X_D = a'b'Y_3'$$

$$X_c = (a + f') \cdot Y_1$$
SET Y_1 = START $\overrightarrow{(\text{DEL } 3.0 \text{ sec.})}$
RESET Y_1 = STOP
SET $Y_2 = e$
RESET Y_2 = RESET $Y_3 = a$
SET $Y_3 = c'$

As the reader has become familiar with the construction of the attached logic diagrams (subject of the previous section), no in-depth discussion is offered relating to the development of the diagram of Figure 4-36. The diagram is shown complete with the necessary accessories, i.e., a filter, a strainer, and a pressure controller. The reader may note the complexity of the diagram when all lines must be connected. It is therefore accepted as a general practice to simply omit the lines that may obstruct the clarity of the diagram and to label the "unconnected" lines properly to indicate the existence of an omitted connection.

The detached logic diagram is constructed in a manner discussed earlier in this section. In the design of the detached diagram of Figure 4-37, the authors have taken the freedom of assigning relay valves to lines 1, 8, 9, and 11, which are actually not absolutely necessary. This step was performed for convenience in constructing the diagram. Observing the diagram shown, the reader will agree that the construction of the detached diagram is more manageable than the attached diagram. However, the use of the diagram may not be appropriate when specially designed logic elements (not formed by three- or four-way valves) are used, as the representation of such elements in the diagram may not be directly apparent.

PROBLEMS

1. Represent the AND, OR, NOT, and NAND operations by the relay-valve symbology.
2. Represent the EXCLUSIVE OR, COINCIDENCE, INHIBITOR, AND/INHIBITOR, and COINCIDENCE/EXCLUSIVE OR elements by the relay-valve symbology.
3. Implement the following equations using any of the logic elements shown in this chapter, and use memory elements where necessary:
    a $Z = AC' + B'CD + A'BD'$
    b $Z = ABC + A'C'D + A'B'D'$
    c $Z = A(C'D + B'D) + A'B'C'D'$
    d $Z = (A' + B'C)(B' + DE + A)$
    e $Z = (B + D)(A + B + C)(A' + B + C')$
    f $Z = AC' \cdot (BD' + B'D) + AC'E$
    g $Z = AB'C' + Z\,ED + ABC + Z(FG' + H)$
4. Implement the equations of Problem 3 by NAND and NOR elements which have
    a a fan-in limit of 2
    b a fan-in limit of 4
    c unlimited fan-in

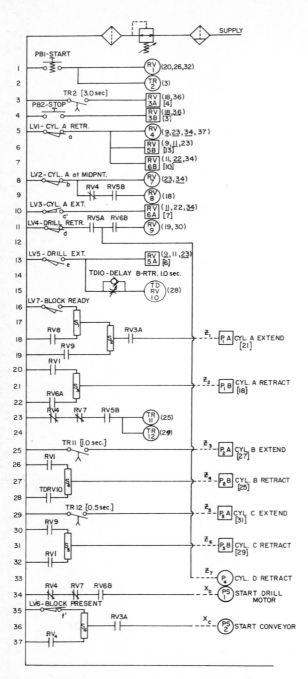

Figure 4-37 A.N.S.I. detached logic diagram.

LOGIC ELEMENTS AND CIRCUITS

5 Implement the following network equations. Use elements that are available and that will minimize the number of elements involved in the circuit.
   a $Z_1 = x'_1 \, x'_2 \, x'_3 \, y'_1$
   b $Z_2 = y_1 y_2 x'_2 + y_1 y'_2 y_3$
   c $Z_3 = y_1 \, y'_2 y'_3 x'_2$
   d $y_1 = y_1 x'_3 + y_1 x_2 + y_2 y'_3 x_1$
   e $y_2 = x_1 + y'_1 y_3 + y'_1 y_2 + y_2 y'_3 + y_2 x_2$
   f $y_3 = x_1 y_3 + x_1 y'_2 + y_1 y_3 x'_2 + x_2 y_2$

6 For the problems given in Problem 5, construct the A.N.S.I.-approved detached relay diagram. Assume that all inputs were generated by mechanical sensors.

7 Illustrate by using the attached and detached diagrams, the implementation of the following network equations:
$$Z_{M1} = A(\overleftarrow{\text{DEL}}\ 1.0 \text{ sec}) + CY_1$$
$$Z_{M2} = DY'_3 + (EY_1 + B)(\overleftarrow{\text{DEL}}\ 0.5 \text{ sec})$$
$$Z_N = CBY'_1 + DBY'_2 Y_3 + EBY'_1$$
$$Z_Q = (DY'_3 + EY_1 + B)(\overrightarrow{\text{DEL}}\ 0.8 \text{ sec})$$
SET $Y_1 = DY'_3$
RESET $Y_1 = DY_2 Y_3$
SET $Y_2 = EY_1$
RESET $Y_2 = FY'_3$
SET $Y_3 = CY_1$
RESET $Y_3 = AY_1 Y'_2$

Signals A, B, C, and D are generated by sensors that are used to detect the positions of cylinders M, N, and the electric motor Q. Input E is a manual input signal. Output signal Z_{M1} and Z_{M2} are the "extend" and "retract" signals for cylinder M, which is actuated by a detented four-way valve. Output Z_N activates a spring-return valve of cylinder N, while Z_Q activates the electric motor Q.

8 The following circuit is to be implemented by a mixed logic circuit with detented output actuator valves:
SET $Y_1 = B$
RESET $Y_1 = C$
SET $Y_2 = A'B'C'Y_1 Y_3$
RESET $Y_2 = BY_3$
SET $Y_3 = A'B'C'Y_1 Y_2$
RESET $Y_3 = C + AY_2$
$Z_A = A'B'C'Y'_1 \, Y'_2 + A'B'C' \, Y' \, Y_3$
$Z'_A = A$
$Z_B = A'B'C'Y_1 Y_2 + A'B'C'Y_2 Y'_3$
$Z'_B = B$
$Z_C = A'B'C'Y_1 Y'_2$
$Z'_C = C$

Draw the attached and the detached diagrams.

9 For the network of Problem 8 it is desired to obtain both NOR and NAND networks. The detented output valves should be replaced by a spring-loaded actuator valve. Draw the network representation. Hint: The detented outputs should be treated as memory devices.

10 Work Problem 7 using NAND and NOR elements.

11 Use the truth table approach to derive the various utilizations of existing three-way or four-way valves (spring-return, spring-centered, open-centered, blocked-centered, etc.). In order to obtain a general solution, connect every input and control port to an individual signal. The "various uses" of the valve can be obtained by connecting one or more ports to either tank 0 or pressure 1.

12 Implement Problems 7 and 8 using only three-way or four-way valves, making use of any of the possible functions that can be achieved by these valves.

DEFINITIONS

Accessory element A device that is not a logic element but is needed to implement a complete logic circuit.
Active element A logic element that has one of its passage ports connected to a power source.
Control point of a valve The actuation port of a valve.
Fan-in The number of permissible logical inputs to a logic element.
Fan-out The number of logic elements of like kind that can be operated or controlled by the given element when all elements are operating at the same pressure.
Flow passage (— of a valve) A path that internally connects one port to another, at one or more, but not all, actuating positions of the valve.
Imperfect logic element A logic device that satisfies a part of the logical relationships stated by the prescribed truth table.
Logic element A device that is capable of making a TRUE or FALSE, YES or NO, "1" or "0" output decision, based upon its input condition.
Passive logic element A logic element that has none of its ports connected to a power source.
Perfect logic element A logic element that satisfies all logical relationships stated by the prescribed truth table.
Prepared path condition A condition where the change of the input causes a desired change of the output without any switching of logic elements.
State One of the distinct conditions of a logic device or signal.
Unary element An element that has one input.

REFERENCES

Auger, R. N., "How to Use Turbulence Amplifiers for Control Logic," *Control Engineering*, pp. 89–93, June, 1964.
Belsterling, C. A., *Fluidics Systems Design*. New York: Wiley Interscience, John Wiley and Sons, Inc., 1971.
Bouteille, D., and C. Guidot, *Fluid Logic Controls and Industrial Automation*. New York: Wiley Interscience, John Wiley and Sons, Inc., 1973.
Brewin, G. M., "Disc Valve Logic Devices," *Instruments and Control Systems*, Vol. 41, pp. 91–93, April 1968.
Cole, J. H., "Synthesis of Optimum Complex Fluid Logic Sequential Circuits," Ph.D. Dissertation, Oklahoma State University, 1968.

DeMoss, D. M., "Criteria For the Design of Fast, Safe, Asynchronous Sequential Fluidic Circuits," Ph.D. Dissertation, Oklahoma State University, 1967.

Dietmeyer, D. D., *Logic Design of Digital Systems*. Boston: Allyn and Bacon, Inc., 1972.

Doig, G., and Walle, L. I., "How to Design All-Air Control Systems for Automatic Machines," *Practical Air Circuitry Manual Pac-65*, Numatics, Inc., Michigan, 1966.

Fitch, E. C., Jr., "Fluid Logic." A Teaching Manual Published by The School of Mechnical and Aerospace Engineering, Oklahoma State University, Oklahoma, 1966.

Iseman, J. M., "The Application of Fluidics to Low Power Logic Circuits," American Society of Mechanical Engineers, Paper No. 68-DE-29, 1968.

Lueth, J. H., "Analogies Between Electrical and Fluid Logic Control Elements," *Hydraulics and Pneumatics*, November Issue, pp. 95–100, 1966.

Surjaatmadja, J. B., "A Generalized Method for Synthesizing Optimal Fluid Logic Networks," Fluid Power Research Center, Oklahoma State University, Oklahoma, Report R73-FL-3, 1973.

Togino, K., and Inoue, K., "Universal Fluid Logic Element," *Control Engineering*, pp. 78–87, May 1965.

———, "Directional Valves for Profit-Making Designs," *Hydraulics and Pneumatics*, pp. HP1–HP32, March 1973.

———, "Fluid Logic," *Machine Design*, Fluid Power Reference Issue (September 12), Section 12, pp. 258–266, 1972.

———, "ARO Pneumatic Logic Controls," Module 1 to 10, The ARO Corp., Bryan, Ohio, 1973.

———, "Method of Diagramming for Moving Parts Fluid Controls," American National Standards Institute, A.N.S.I./B93.38, 1976.

Chapter 5

Logic System Description

5-1 FUNDAMENTAL CONSIDERATION

The design of a logic system begins with a statement of the problem or the expected machine operation. Basically, this involves an explicit formulation of the conditions which must exist before each and every output is activated. Such a formulation is a firm set of operational specifications which is logically meaningful and presented in a manner that is unambiguous and noncontradictory. This means that simple statements are needed covering every combination of logical conditions as to what is required, what is not desired, and what is of no concern or importance.

The format used to describe the requirements of a logic system can vary somewhat in complexity for the type of network involved. In every case, however, the description must establish the logic program for the output signals. For a combinational logic network where an output is required when a specific combination of input signals exists, the descriptive format can be much simpler than for a sequential-type system where the output is a result of not just one combination of inputs but a series of inputs. Even for the case of the sequential-type networks, some descriptive formats are not suited for general use. For example, some formats can satisfy deterministic-type sequential-logic networks, but are incapable of

LOGIC SYSTEM DESCRIPTION

expressing a stochastic-type network. Definitions for these sequential type networks are as follows:

Deterministic logic circuit A sequential network where both the sequence of inputs and the machine operation are predetermined.
Stochastic logic circuit A sequential network where predetermined, random, or both types of inputs are permissible.

Regardless of the type of network being designed, a method must be used for insuring that all possible combinations of input states affecting the operation are considered in prescribing the output conditions. The formatting techniques presented and used in this chapter represent the most effective means available for describing the requirements of a logic system.

5-2 TRUTH TABLE

The truth table is undoubtedly the earliest type of logic system description. Since the time factor or the history of inputs are not considered in the description, truth tables are only applicable for pure combinational networks—those where the output of the logic system is solely dependent upon the state of the inputs. For these systems, the output state must always be the same for a particular state of the inputs. To illustrate the use of the truth table, consider a control network which is required to perform the sorting of a batch of boxes from a conveyor as shown in Figure 5-1. The conveyor on the right transfers boxes of three different sizes, which can be classified as small, medium, or large. The logic network is expected to control the following marshaling schedule:

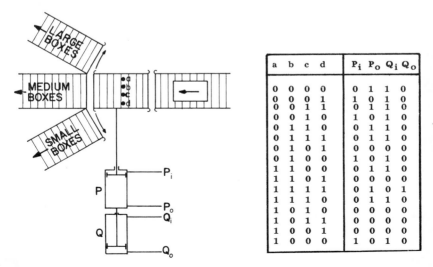

a	b	c	d	P_i	P_o	Q_i	Q_o
0	0	0	0	0	1	1	0
0	0	0	1	1	0	1	0
0	0	1	1	0	1	1	0
0	0	1	0	1	0	1	0
0	1	1	0	0	1	1	0
0	1	1	1	0	1	1	0
0	1	0	1	0	0	0	0
0	1	0	0	1	0	1	0
1	1	0	0	0	1	1	0
1	1	0	1	0	0	0	0
1	1	1	1	0	1	0	1
1	1	1	0	0	1	1	0
1	0	1	0	0	0	0	0
1	0	1	1	0	0	0	0
1	0	0	1	0	0	0	0
1	0	0	0	1	0	1	0

Figure 5-1 Box-sorting from a conveyor and the truth table.

1 Send the large boxes to the conveyor at the top.
2 Send the medium-sized boxes to the middle conveyor.
3 Send the small boxes to the bottom conveyor.

There are various design approaches for the input sensors that can be used to perform the prescribed task. One approach is shown in Figure 5-1, where a four-sensor pattern is utilized to identify box sizes. The spacing of the sensors is such that when a small box passes the sensing zone, only one sensor is activated. Two or three activated sensors indicate that a medium box is passing the zone, while a large box is indicated by inputs from all the sensors. The box-diverting function of the conveyor is performed as follows: boxes are routed to the top conveyor by extending both cylinders P and Q, to the middle conveyor by retracting one cylinder, and to the bottom conveyor by retracting both cylinders.

The logic specification statement for the box-marshaling problem can be formally written as follows:

1 Extend both cylinders ($P_o = Q_o = 1$) when all inputs are activated.
2 Extend cylinder P and retract cylinder Q when two or three adjacent input sensors are ON ($P_o = Q_i = 1$).
3 Retract both cylinders when only one of the sensors is activated ($P_i = Q_i = 1$).
4 Extend cylinder P and retract cylinder Q if otherwise ($P_o = Q_i = 1$).

Since no time dependency is implied or exists in the above specification, a truth table can be used to describe the logic of the system.

The rules for constructing a truth table for a combinational-logic system are simple and can be listed as follows:

1 List all possible input states on the left side of the table.
2 Using the logic specification of the system, record the value of the required output state for each combination of input states.

The complete truth table for the box-marshaling problem is shown in Figure 5-1.

5-3 TIMING OR BAR CHART

When the time factor becomes involved in a logic specification, truth tables cannot be utilized for the problem description. A slight modification of the truth table by allowing the time factor to become one of the variables of the descriptive format results in the formulation of a "timing chart" or a "bar chart." Initially, bars were used instead of binary numbers to represent the presence of an "active" "ON condition" or a Boolean "1". It is generally accepted that when binary values are used in the chart it is referred to as a "timing chart" and when bars are used to reflect periods of activation the chart is called a "bar chart."

The timing or bar chart is a chart which is capable of keeping track of each operation with respect to time. The term "time" here is used to represent relative

LOGIC SYSTEM DESCRIPTION

time with respect to a certain machine cycle and should not be confused with an absolute time scale. The timing chart is a particularly popular means for describing the sequence for deterministic processes (both combinational and sequential) and is used in areas other than logic, where accounting of processes is utilized (e.g., in process control, scheduling of operations or operators in a certain plant, etc.). The format of the timing chart resembles the chart utilized in scheduling or dispatching products for industrial control. At the top of the chart, time is represented numerically by the various operational time periods of the machine cycle. Each period is assigned a separate column in the chart, while each row is dedicated to a particular machine variable. Bars or binary numbers are used to represent the activation of the particular variable in the associated time period. Of course, the bar representation is changed to binary numbers when computers are used in the design process.

To illustrate the timing and bar chart technique, consider again the conveyor set-up shown in Figure 5-1. Assume that the task now involves sequential time and can be expressed by the following operational sequence:

1 Send three boxes to the top conveyor.
2 Send two boxes to the bottom conveyor.
3 Send one box to the top conveyor.
4 Send four boxes to the middle conveyor.
5 Repeat the cycle.

Since only the presence of a box must be detected, it is assumed that a sensor a is employed which will be tripped every time a box passes the location a regardless of the size of the box. It should be apparent that this task will require a deterministic-type sequential-control network. The logic specification is therefore:

1 Extend both cylinders when sensor a is activated. Remain extended during three complete cycles of input signal a (where input signal a is the input signal produced by sensor a).
2 Retract both cylinders when sensor a is activated and remain retracted during two complete cycles of input signal a.
3 Extend both cylinders during one cycle of input signal a.
4 Retract cylinder Q during four consecutive cycles of input signal a.

The timing or bar chart provides a straightforward information source concerning the condition of every input, output, and possible memory requirement with respect to every accountable instant of time. The chart can be constructed using the following rules:

1 Assume an initial condition for the machine cycle (a good condition might be where the first operation is to begin). Record this initial condition (input and output) in the time period or state 1.
2 Record the input state that will occur next in accordance with the specification and the configuration of the machine.

Table 5-1 The Bar Chart

Variable	Time Period/States																				
	1	2	3	4	5	6	7	8	9	10	11	12	13	14	15	16	17	18	19	20	1
a																					
P_o																					
P_i																					
Q_o																					
Q_i																					

3 Assign the appropriate output as specified by the logic specification.
4 Repeat Steps 2 and 3 until the cycle reaches the previously determined initial condition.

The bar chart for the box-dispatching problem can be constructed as shown in Table 5-1, while its numerical version is as displayed in Table 5-2. The initial condition used in the construction of both charts was assumed to be at the point where a box was just sensed (a is on) and that three boxes are to be dispatched to the top conveyor; hence, P_o and Q_o are both activated during three full cycles of a (until machine state 6 is reached). Within time period 7 through 9, two boxes are sent to the bottom conveyor. State 20 is the last state in the cycle and the state where the machine returns to its original state. It should be noted that each time

Table 5-2 The Timing Chart

Variable	Time Period/States																				
	1	2	3	4	5	6	7	8	9	10	11	12	13	14	15	16	17	18	19	20	1
a	1	0	1	0	1	0	1	0	1	0	1	0	1	0	1	0	1	0	1	0	1
P_o	1	1	1	1	1	1	0	0	0	0	1	1	1	1	1	1	1	1	1	1	1
P_i	0	0	0	0	0	0	1	1	1	1	0	0	0	0	0	0	0	0	0	0	0
Q_o	1	1	1	1	1	1	0	0	0	0	1	1	0	0	0	0	0	0	0	0	1
Q_i	0	0	0	0	0	0	1	1	1	1	0	0	1	1	1	1	1	1	1	1	0

period actually represents one machine state; hence, the use of the term "timing" may not seem appropriate.

The timing-chart concept applies to a combinational-type problem where a sequence of machine timing is involved. In the illustrative problem presented in Section 5-2, no such schedule of machine-operational events was implied; hence, there is no need to resort to any method of describing the logic system other than the truth table itself. Since most machines have a "preferred" sequence of operations, the timing chart or the bar chart has an important role in fluid logic design.

5-4 THE OPERATIONS TABLE

The operations-table method of describing the word statement of a logic problem was advanced by J. H. Cole. This approach was developed for systems which were "regularly activated" or where the sequence of operation is deterministic in nature. Although the approach is applicable to all deterministic-type problems, it was originally designed to serve as a synthesis format for a very particular class of logic systems. The following restrictions must be placed upon the operation of the system:

1. It must have a distinct, known sequence of operation.
2. The inputs must serve as a feedback to indicate whether an operation has been completed.
3. Only one operation may be performed at a time.

With the above restrictions, only the changed signals can actually influence the activation of an output of the system. Therefore, it is the primary objective of the operations table to reveal the newly changed signal resulting from the finalization of the previous operations and to record the signal changing in the table for each machine state.

The operations table can be constructed by applying the following steps:

1. List the sequence of operations in the first column. This is accomplished by considering the operation "cylinder A extends" as being indicated by simply A, while its retraction is indicated by $\bar{A}$.
2. By inspection, list the input that is changed as a result of the last operation. This operation is performed for all operations of the system.
3. If desired, the input states can be determined by using the same approach, leaving an input unchanged until it is changed by some particular operation.

In order to illustrate the operations-table method of describing a logic problem, consider a two-cylinder, two-input problem that is expected to control the following sequence: A, $\bar{A}$, B, $\bar{B}$, B, A, $\bar{A}$, $\bar{B}$, A, $\bar{A}$, and the cycle repeats. Assume also that the completion of the operation A is indicated by an input signal a while an a' indicates the completion of the retraction operation.

The operations table can be constructed by initially placing the complete operation of the machine in the second column of the table (see Table 5-3). The

Table 5-3 An Operations Table

State No.	Operation	Inputs	Changed Inputs	Memory Assignment and Excitation
1	A	$a'b'$	a'	
2	$\overline{A}$	$a\,b'$	a	
3	B	$a'b'$	a'	
4	$\overline{B}$	$a'b$	b	
5	B	$a'b'$	b'	
6	A	$a'b$	b	
7	$\overline{A}$	$a\,b$	a	
8	$\overline{B}$	$a'b$	a'	
9	A	$a'b'$	b'	
10	$\overline{A}$	$a\,b'$	a	

first operation (at state 1) is operation A. Hence, the changed state of the next state (state 2) is a, which is also the state of the input variable a (in columns 3 and 4). It can be noted, at this time, the condition of signal b is still unknown, which is revealed only after reaching state 4. At this point, cylinder B has been extended and signal b is activated, while signal a remains unchanged. This procedure is continued until all conditions of the inputs are considered.

The simplicity of the operations table is quite apparent. Although no stochastic problems can be specified by the operations table, this format is definitely convenient to use when only deterministic (regularly activated) networks are encountered.

5-5 PRIMITIVE FLOW TABLE

Soon after Karnaugh developed his now famous map method for combinational logic synthesis, Huffman formalized a technique for deriving sequential network equations utilizing the properties of Karnaugh maps. A major breakthrough which must be credited to Huffman is the advancement of the Primitive Flow Table (PFT) method of describing a sequential logic system specification. This flow table provides a means of recording the "history of inputs."

The PFT is constructed from the word statement of the problem and exhibits the desired operational sequence in accordance with both input and output conditions. This rudimentary table contains 2^n columns to represent every combination of the n inputs. These columns are identified by the Gray Code as used in the Karnaugh maps. A separate row is used in the table for each stable machine state. On the right side of the table are listed the outputs of the system which are associated with each machine state. In the cells formed by the intersections of the rows and columns, both stable and unstable state conditions of the memory are

LOGIC SYSTEM DESCRIPTION

recorded. A stable state is indicated by bracketed, parenthetical, or circled entries (see Table 5-4), while plain characters denote unstable states. A stable state refers to a condition of the memory in which no change of the machine state can occur without some change in the input state. An unstable state indicates a transitional state where the plain character used refers to the "next" state that should occur due to a change of the inputs. The concept of stable and unstable memory-state conditions is most important in the construction of the PFT and a deeper understanding of their meaning is deemed necessary.

Basically, the stability characteristics of an element or circuit are associated with the time delay which always accompanies the time a memory is signaled to change and the time when the change is finalized. If the memory is represented by y and its excitation signal by Y, then the system is stable when the value of y is the same as Y or is unstable when the excitation signal does not agree with the state of the memory. As an example, for a system containing one memory (y), if $Y = 1$ and $y = 1$, or $Y = 0$ and $y = 0$, then the system is stable; but if $Y = 1$ and $y = 0$, or $Y = 0$, and $y = 1$, then the memory is unstable. The next state of a memory system must always agree with its present excitation. For example, if $Y = 1$ and $y = 0$, then the signal-element memory is unstable and the state of the memory must eventually change to $y = 1$ to produce a stable condition. Of course this concept of stability applies equally well to multielement memory systems.

The rows of a PFT can be considered as representing the various machine states, while the entries in the cells of the table are the states of the memory. Since only one stable memory state is permitted per row in the Primitive Flow Table, a stable condition exists when the memory state is numerically equal to the machine state. A change in the input "triggers" the memory to change to another machine state. In other words, when an input change occurs, the network moves from a stable state to an unstable state in the same row and then transfers to a stable state in another row.

Table 5-4 The Primitive Flow Format

NEXT STATES					OUTPUT STATES
	INPUT STATES $f(a,b{-}{-}{-})$				$Z_1 Z_2 {-} {-} Z_n$
f_1	$f_2 {-}{-}{-}{-}{-}{-}{-} f_i {-}{-}{-}{-}{-}{-}{-}{-} f_n$				
$\gamma \leftarrow [\alpha] \longrightarrow \beta$					$g_\alpha(Z_1 {-}{-}{-} Z_n)$
γ			[δ]		$g_\delta(Z_1 {-}{-}{-} Z_n)$
	[β] $\longrightarrow \delta$				$g_\beta(Z_1 {-}{-}{-} Z)$
[γ]					$g_\gamma(Z_1 {-}{-}{-} Z_n)$

The mechanics associated with the state changes in a PFT can be illustrated using Table 5-4. It can be noted that a change in the input state is always required to initiate a change in the machine state, and this change is indicated by a horizontal transfer in the PFT followed by a vertical transfer to the associated stable state in the same column (but in a different row). As an example, if the system is initially at state α (with an input state of f_2), and the input f_2 changes to f_i, the system will find an unstable state β in the same row corresponding to the input f_i, then it will seek its related stable state β by transferring vertically in the f_i column.

The outputs (or output states) are listed on the right side of the PFT. If all the output states represent the outputs of the stable states, then, at most only one output state exists for each row. In this case, the output states can be represented in one column, as shown in Table 5-4. However, if it is required to determine the outputs of one or more unstable machine states, the expansion of the output subtable (that part of the table describing the output relations) is necessary. This is performed by listing all the input states in a similar fashion as on the left side of the table (the left portion of the PFT is often referred to as the "next state subtable"). In Table 5-5, such a situation is illustrated. Note that every position in the output subtable corresponds to a related machine state in the next state subtable. For example, $g_{\beta u}$ is the output state of the machine while it is in its unstable state β, etc. It might be worth mentioning that in fluid logic networks, it is seldom necessary to consider the outputs of an unstable state, and therefore the format of the PFT as given in Table 5-4 is the one most often utilized.

The actual construction of the PFT is simple and can be made directly from information given in the problem specification. The following rules should be applied:

1 Determine the number of input variables, *e.g.*, n. Hence, there will be 2^n columns in the next state subtable (the main left portion of the table).

Table 5-5 The Primitive Flow Table Format with an Expanded Output Subtable

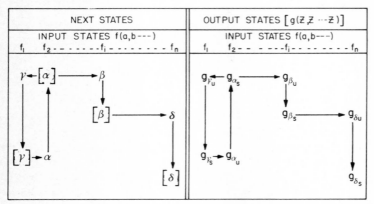

LOGIC SYSTEM DESCRIPTION

2 Determine from the logic specification whether or not the outputs are specified for the unstable states. If they are required, provide 2^n columns for the output subtable; otherwise, provide only one column for the output state.
3 Code all possible input states using the Gray Code format and list them in the columns of the next state subtable. Also, list them in the output subtable if it is in the expanded form.
4 Select an initial condition in the cycle. Assign a character to the designated initial state and place it in the first row of the column associated with the appropriate input state. Indicate its stability by brackets, parentheses, or circle.
5 In the appropriate location in the output subtable (in the same row), assign the required output state for that particular state.
6 Determine all possible input states that may occur "next" in relation to the logic specification of the problem. In each of the associated column(s), in the same row, characters are assigned to indicate the next-state conditions of the machine. Note that in this same row, these next states must always be unstable.
7 If specific output states are required for any unstable states, then the output states are assigned to each related location in the expanded output subtable.
8 Select a character that has not been used to designate the next stable state. In a new row, enter this character with brackets to indicate its stable condition. Perform Steps 5 through 8 iteratively until all possible states of the machine are contained in the flow table.

In order to illustrate the procedure for constructing a Primitive Flow Table, consider the following example problem:

The conveyor in Figure 5-1 is supposed to send one large box to the top conveyor, then send a medium box to the middle conveyor, directly followed by a small box. All other boxes that do not satisfy the required sequence are to be sent to the lower conveyor. Boxes on the conveyor are originally placed in a random manner.

To accomplish the box identification needed for this system, the alignment of the sensors as shown in Figure 5-2 can be used. In this case, the sensors are tripped in a sequential manner, depending upon the existence of three possible situations: *(1)* all three signals energized simultaneously indicates a large box, *(2)* two coincident signals at any time indicates a medium-size box, *(3)* a small box is detected by a one-at-a-time activation of the three sensors. As there are three inputs available from the sensor system, 2^3 or 8 columns are needed to represent all possible input states, and these are coded systematically and shown at the top of these columns. Assume that the initial state of the system is the point where a large box is to be sent to the top conveyor and label this as state (1). At this particular state, it can be also assumed that the moving conveyor is being directed to the bottom conveyor ($P_i = Q_i = 1$).

The first and only change of signal which must occur is the activation of the first sensor a. This change in system input should cause a transfer of the network to the next state, which is labeled as state 2. At this stage, no decisions can be made, as the size of the box passing is still unknown. Hence, it is practical to leave both

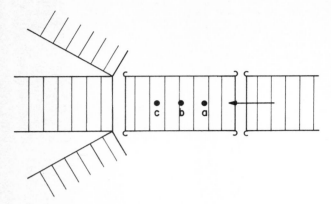

Figure 5-2 In-line sensor configuration.

cylinders P and Q in their fully retracted positions. It is known, however, that there are two possible input states that may occur next—the activation of sensor b, which means that the box is either a medium- or large-size box, or the deactivation of sensor a, which would indicate that the box passing was a small-size box. The two possible next states are indicated by states 3 and 4, respectively. Similarly, from state 3, two other possibilities exist: one that the box is medium size (by the deactivation of signal a), and the second that a large box exists (by the activation of signal c). On the other hand, there is only one possibility that can occur following state 4, and that is since it has already been determined that the box passing is definitely a small box, state 4 must be followed by a successive activation and deactivation of signals b and c. The completed Primitive Flow Table is illustrated in Table 5-6. It should be obvious to the reader that there is no way for this stochastic-type sequential machine problem to be expressed by either a truth table or a timing chart.

5-6 THE STATE MATRIX

A somewhat different format for describing a word statement of a logic problem was advanced by R. L. Woods. The format is known as the state matrix and assumes that all problems can be described by the matrix representation

$$[Z] = [M] \cdot [X]$$

where Z is the output matrix, M the network matrix, and X the input matrix. The Z matrix is a matrix formed by all the outputs and their complements, while the input matrix X contains all the input states of the system listed in the Gray Code format.

Every row of the M or network matrix is dedicated to a particular output or its complement, while every column is related to a particular input state. A state is entered in every position of the network matrix that satisfies the input-output

LOGIC SYSTEM DESCRIPTION

relationship. The procedure for constructing the matrix is as follows:

1. Select an initial condition and arbitrarily assign a state number, e.g., state 1. This state number is entered in every matrix location under the column related to the input state, in row locations corresponding to the output conditions.
2. Determine the next possible states of state 1 and place them in appropriate locations in the M matrix considering their inputs and outputs.
3. As a supplement of the Woods method (to enhance the matrix method as a problem-description technique), place in parentheses beside each state in the M matrix the identification of the next state.

To illustrate the state matrix, consider the example problem used to describe the Primitive Flow Table method. The resulting matrix representation is illustrated in Figure 5-3. Note that since P_i is the complement of P_o, and Q_i is the complement of Q_o, the complements of these outputs need not be presented in the matrix.

Table 5-6 The Primitive Flow Table

	NEXT STATES							OUTPUT STATES			
abc 000	001	011	010	110	111	101	100	P_o	P_i	Q_o	Q_i
|1|							2	0	1	0	1
4				3			[2]	0	1	0	1
			6	|3|	5			0	1	0	1
|4|			7					0	1	0	1
		8			[5]			1	0	1	0
		9	|6|					0	1	0	1
10			|7|					0	1	0	1
	11	|8|						1	0	1	0
	12	|9|						0	1	0	1
|10|	13							0	1	0	1
14	|11|							1	0	1	0
1	|12|							0	1	0	1
1	|13|							0	1	0	1
|14|						15		0	1	0	1
17				16		[15]		0	1	0	1
			19	|16|	18			0	1	0	1
|17|			20					0	1	0	1
		21			[18]			0	1	0	1
		22	|19|					0	1	0	1
23			|20|					0	1	1	0
	24	|21|						0	1	0	1
	25	|22|						0	1	1	0
|23|	26							0	1	0	1
14	|24|							0	1	0	1
27	|25|							0	1	1	0
14	|26|							0	1	0	1
|27|							·28	0	1	0	1
30				29			[28]	0	1	0	1
			32	|29|	31			0	1	0	1
|30|			33					0	1	1	0
		34			[31]			0	1	0	1
		35	|32|					0	1	0	1
36			|33|					0	1	1	0
	37	|34|						0	1	0	1
	38	|35|						0	1	0	1
|36|	39							0	1	1	0
27	|37|							0	1	0	1
27	|38|							0	1	0	1
1	|39|							0	1	1	0

(Left margin brackets: Large Box to Top Conv.; Medium Box to Middle Conv.; Small Box to Middle Conv.)

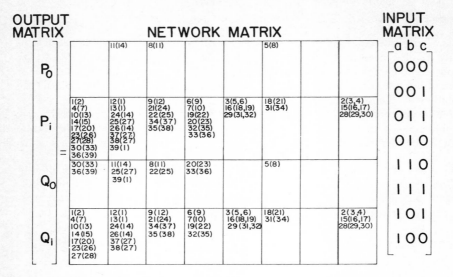

Figure 5-3 Illustration of the state matrix.

5-7 THE SYNTHESIS TABLE

In order to consider irregularly activated stochastic networks, G. E. Maroney developed the "Diconesyn III Synthesis Technique." To achieve his objectives, Maroney promoted the use of a special type of problem description which he termed the Synthesis Table. No restrictions are made with respect to the type of problem that can be recorded by this format. The method of construction of the table is similar to the construction of the Primitive Flow Table, with the exception that the machine states and the corresponding input states are listed row-by-row along with the "next possible states." The procedure can be written as follows:

1. Select an initial condition and arbitrarily assign a state number, e.g., state 1, which is entered in column I.
2. Record the outputs relating to the individual states in column II.
3. Determine the associated input state and place the values in column III.
4. Determine the next possible state of state 1 and arbitrarily designate it as state 2. This state is placed in row 2 of the table and the inputs and outputs are recorded as indicated in Steps 1, 2, and 3. In addition, a "2" is placed in row 1, column IV, to indicate that state 2 is the next possible state of state 1.
5. Check for other possible next states of state 1, if any. These are placed in subsequent locations in the table and are further identified as in Step 4. If no such case exists, continue to consider the next state (states 2, 3, etc.) and return to Step 4. As in the Primitive Flow Table approach, note that the next state may be a state that has been considered previously.
6. Perform Steps 3 and 4 until no other new states exist.
7. Determine the previous state(s) (the one(s) directly preceding the given state) for each machine state and record them in column V.

LOGIC SYSTEM DESCRIPTION

In order to illustrate the Synthesis Table, consider the problem description used for the Primitive Flow Table case. The result can be observed in Table 5-7.

5-8 THE LOGIC SPECIFICATION CHART

As the complexity of the logic statement increases, the descriptive formats which have been presented become untenable. A method must be employed which can

Table 5-7 The Synthesis Table

State no.	Outputs $P_o\ P_i\ Q_o\ Q_i$	Input state	Next states	Previous states	Synthesis
1	0 1 0 1	$a'b'c'$	2	12, 13, 39	
2	0 1 0 1	$ab'c'$	3, 4	1	
3	0 1 0 1	abc'	5, 6	2	
4	0 1 0 1	$a'b'c'$	7	2	
5	1 0 1 0	abc	8	3	
6	0 1 0 1	$a'bc'$	9	3	
7	0 1 0 1	$a'bc'$	10	4	
8	1 0 1 0	$a'bc$	11	5	
9	0 1 0 1	$a'bc$	12	6	
10	0 1 0 1	$a'b'c'$	13	7	
11	1 0 1 0	$a'b'c$	14	8	
12	0 1 0 1	$a'b'c$	1	9	
13	0 1 0 1	$a'b'c$	1	10	
14	0 1 0 1	$a'b'c'$	15	11, 24, 26	
15	0 1 0 1	$ab'c'$	16, 17	14	
16	0 1 0 1	abc'	18, 19	15	
17	0 1 0 1	$a'b'c'$	20	15	
18	0 1 0 1	abc	21	16	
19	0 1 0 1	$a'bc'$	22	16	
20	0 1 1 0	$a'bc'$	23	17	
21	0 1 0 1	$a'bc$	24	18	
22	0 1 1 0	$a'bc$	25	19	
23	0 1 0 1	$a'b'c'$	26	20	
24	0 1 0 1	$a'b'c$	14	21	
25	0 1 1 0	$a'b'c$	27	22	
26	0 1 0 1	$a'b'c$	14	23	
27	0 1 0 1	$a'b'c'$	28	25, 37, 38	
28	0 1 0 1	$ab'c'$	29, 30	27	
29	0 1 0 1	abc'	31, 32	28	
30	0 1 1 0	$a'b'c'$	33	28	
31	0 1 0 1	abc	34	29	
32	0 1 0 1	$a'bc'$	35	29	
33	0 1 1 0	$a'bc'$	36	30	
34	0 1 0 1	$a'bc$	37	31	
35	0 1 0 1	$a'bc$	38	32	
36	0 1 1 0	$a'b'c'$	39	33	
37	0 1 0 1	$a'b'c$	27	34	
38	0 1 0 1	$a'b'c$	27	35	
39	0 1 1 0	$a'b'c$	1	36	

suitably describe the logic of a large-scale problem and be adaptable to computer use. The Logic Specification Chart (LSC) incorporates the "best" features of the other methods and satisfies the following objectives:

- Simple to construct and understand.
- Complete and unambiguous.
- Able to describe any finite-state, large-scale sequential system.

One of the main features of the chart is that it minimizes the number of input columns by disregarding the unused input states.

	c	0	0	0	1	0	1	1
	b	0	0	1	1	1	1	0
	a	0	1	1	1	0	0	0
OUTPUTS		INPUT STATES
$P_i\ P_o\ Q_i\ Q_o$		I_1	I_2	I_3	I_4	I_5	I_6	I_7
1 0 1 0		‖1‖	2					
1 0 1 0		4	‖2‖	3				
1 0 1 0				‖3‖	5		6	
1 0 1 0		‖4‖					7	
0 1 0 1						‖5‖		8
1 0 1 0							‖6‖	9
1 0 1 0		10					‖7‖	
0 1 0 1							‖8‖	11
1 0 1 0							‖9‖	12
1 0 1 0		‖10‖						13
0 1 0 1		14						‖11‖
1 0 1 0		1						‖12‖
1 0 1 0		1						‖13‖
1 0 1 0		‖14‖	15					
1 0 1 0		17	‖15‖	16				
1 0 1 0				‖16‖	18	19		
1 0 1 0		‖17‖				20		
1 0 1 0					‖18‖		21	
1 0 1 0						‖19‖	22	
1 0 1 0						‖20‖		
0 1 1 0		23					‖21‖	24
1 0 1 0							‖22‖	25
0 1 1 0								26
1 0 1 0		‖23‖						
1 0 1 0		14						‖24‖
0 1 1 0		27						‖25‖
1 0 1 0		14						‖26‖
1 0 1 0		‖27‖	28					
1 0 1 0		30	‖28‖	29				
1 0 1 0				‖29‖	31	32		
0 1 1 0		‖30‖				33		
1 0 1 0					‖31‖		34	
1 0 1 0						‖32‖	35	
0 1 1 0		36				‖33‖		
1 0 1 0							‖34‖	37
1 0 1 0							‖35‖	38
0 1 1 0		‖36‖						39
1 0 1 0		27						‖37‖
1 0 1 0		27						‖38‖
0 1 1 0		1						‖39‖

Figure 5-4 The logic specification chart.

LOGIC SYSTEM DESCRIPTION   129

An example of the Logic Specification Chart is shown in Figure 5-4. Note that the outputs (or the output subtable) are placed on the left side of the table (in reverse order), thus permitting an uninhibited expansion of the "next state" subtable to the right. The LSC can be constructed in a fashion similar to that of the Primitive Flow Table with the exception of the random assignment of the input states. The following steps can be applied in constructing an LSC for a network having outputs assigned to only the stable machine states:

1. Select an initial condition, assign a character for the memory state, and place this character in the first row and the first column of the chart. Label this column with the input state of this initial machine state and indicate the stability of the state by the use of brackets.
2. In the same row, assign the output state of the system at the particular machine state.
3. Determine all of the next possible states as specified by the logic specification of the problem. Assign a different plain character to each of the "next" states and place them in the same row but in the column that has been properly labeled with the associated input state. If no such column exists, assign the next successive empty column to this input state.
4. In each of the columns where a plain character (unstable state) was entered, assign its corresponding bracketed character (stable state) in the same column but in a new row.
5. Repeat Steps 2 through 4 iteratively until all possible states of the machine as dictated by the problem specification have been recorded in the chart.

As an example of this procedure, consider the stochastic problem used to illustrate the Primitive Flow Table. The initial state is selected in the same manner as accomplished for Table 5-6 to enable the reader to compare the two resulting tables. This state is assigned a number 1 and is placed in the first row, first column of the chart, which in turn is assigned the input state of state 1. Stability is indicated by brackets as illustrated in Figure 5-4. The next possible state, state 2 has the input state of $abc = 100$. Since no column has been assigned this particular input, state 2 is placed in the next unassigned column. Like the Primitive Flow Table, the output states are listed in the same row; however, this time on the left side of the chart. As there is only one unstable state that has not yet been assigned a stable state (state 2), the state is entered in the second column, second row of the table. As previously, the next possible states are determined and labeled as states 3 and 4 in Figure 5-4. Only state 3 is assigned a new column, as its input state has not yet been utilized. It should be observed that the completed chart has only seven columns and no unused columns are contained in the chart.

5-9 CONSTRUCTION MECHANICS

In the construction of logic descriptions, there are application details which need to be considered. These details may be classified into three major categories:

130 INTRODUCTION TO FLUID LOGIC

1 Method of starting and resetting the system, including irregular situations such as after power failure or system breakdown.
2 Means for incorporating "nonlogic" elements to satisfy the network requirements.
3 Technique for achieving simplification by utilizing "don't care" outputs.

There are a number of questions which should be answered before attempting to augment the network with the features of category 1. These are:

1 How often do you expect to have a power failure? What action should be taken by the machine during or after such failure?
2 What type of starting mechanisms are desired (ON-OFF switch, ON-OFF dual, or single push button)?
3 Should the machine always return to its initial condition when an OFF signal is sensed?
4 Is the machine designed to be foolproof, i.e., designed to tolerate any irregular conditions? If not, what kind of operator response is desired?

Of course, there is an endless number of combinations of answers to these questions, which oftentimes depends upon the designer. Obviously, answers that strive to minimize resetting (automatically) in the machine usually result in less complex networks. For example, a network will be less complex when an operation is allowed to resume after power shutdown rather than being automatically reset prior to resumption. As an illustration, consider the two-cylinder system shown in Figure 5-5, which has a sequence as follows:

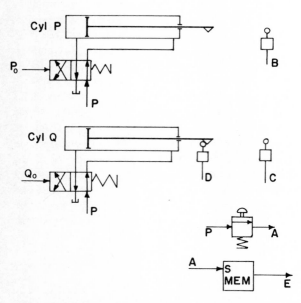

Figure 5-5 Input-output configuration.

LOGIC SYSTEM DESCRIPTION

INPUTS	D	1	1	1	1	0	0
	C	0	0	0	0	0	1
	B	0	0	1	1	1	1
	A	0	1	1	0	0	0

OUTPUTS							
Q_o	P_o	I_1	I_2	I_3	I_4	I_5	I_6
0	0	(1) 2					
0	1	3	[2]	4			
0	1	[3]			5		
1	1				[4] 5		
1	1					[5] 6	
1	1						[6] 7
0	1					8	[7]
0	1				9	[8]	
0	0	1			[9]		

Figure 5-6 LSC of the two-cylinder operation.

1 Following the start signal (signal A), cylinder P is activated.
2 As soon as signal B is energized, cylinder Q is extended until signal C is activated and then is retracted until the sensor for signal D is activated.
3 Retract cylinder P.

The conventional LSC for this sequence is shown in Figure 5-6. Note that no provision is made for any irregular conditions. If it is necessary for the machine to be reset when a power failure occurs, then some type of power failure sensitive device must be used. The biased fluid memory element depicted in Figure 5-5 can serve this purpose. The output from this device (signal E) is utilized in the modified LSC as shown in Figure 5-7. Note the increase in complexity of the LSC, which reflects the degree of complexity of the network.

Generally, an ON-OFF feature to a machine circuit can be added easily by intuition. For example, if the circuit is to stop instantaneously following a stop signal, one may connect this stop signal to the main power valve. On the other hand, if the system is to complete the cycle following an interruption, the outputs (both primary and secondary) at the initial state (causing the change of the state of the machine) should be prevented from occurring. Two illustrations of ON-OFF push button control are presented in Figure 5-8.

A machine network can be made foolproof by considering every possible machine-state condition and designating the desired output. When this is not feasible, a means of overriding all memory and output state conditions is necessary. A typical overriding configuration for a memory or output element is shown in Figure 5-9.

The second category concerns the inclusion of nonlogic devices in a network. Nonlogic devices are incorporated in logic circuits for simplification purposes or for special tasks. These devices include timers, one-shots, amplifiers, and counters. As long as these devices are utilized as input or output devices, their logical representations do not create any major difficulty. However, if such devices are used in the

INPUTS												
	E	0	0	1	1	1	1	1	0	0	0	1
	D	1	1	1	1	1	0	0	1	0	0	1
	C	0	0	0	0	0	0	1	0	0	1	0
	B	0	0	0	1	1	1	1	1	1	1	0
	A	0	1	1	1	0	0	0	0	0	0	0

OUTPUTS													
Q_o	P_o	I_1	I_2	I_3	I_4	I_5	I_6	I_7	I_8	I_9	I_{10}	I_{11}
0	0	(1)	2									
0	1		(2)	3								
0	1			(3)		5					4	
0	1										(4)	
1	1						6					
1	1					(5)	6					
1	1						(6)	7		12		
0	1							(7)	8		13	
0	1								9	(8)		14
0	0								10	(9)	13	
0	0								(10)		12	11
0	0	1		3								(11)
0	1	1								[12]		
0	1									12 [13]	13 [14]	

Figure 5-7 LSC of the two-cylinder operation with provision for power failure.

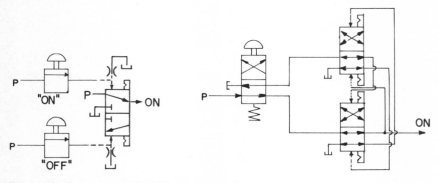

Figure 5-8 Examples of ON-OFF control.

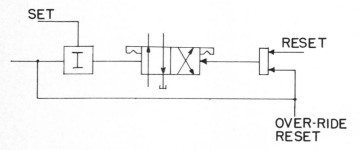

Figure 5-9 A typical override configuration.

LOGIC SYSTEM DESCRIPTION

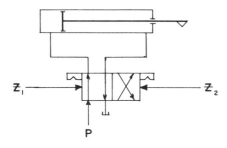

Figure 5-10 A typical output configuration.

control network to perform some logical function, they should be separated from the rest of the logic network. Their inputs can be the outputs of the logic network, while their outputs are fed back to the logic-control portion of the system.

Classified under the last category is the effective utilization of "don't care" outputs. Such "don't care" outputs can originate from a network having an output circuit shown in Figure 5-10. Note that when output Z_1 activates the detented valve shown, then during the next states prior to the activation of output Z_2 the existence of Z_1 will have no influence on the actual operation of the machine. Under this circumstance, the designer should refrain from defining the output state for Z_1, since leaving it as a "don't care" (or unknown) may result in simplifying the network.

PROBLEMS

1 A control system has two inputs, A and B, and two outputs, Z_1 and Z_2, and operates according to the following requirements:
  a Output Z_1 is ON whenever A and B are ON together, and when A and B are both OFF.
  b Output Z_2 is ON whenever A or B is ON.
  Develop the simplest logic system description for representing the problem.

2 A two-cylinder feedback logic control circuit has three inputs, a, b, and c, and four outputs, A_1, A_2, B_1 and B_2. Input signal a is the extension signal for cylinder A, while input b denotes the extension of cylinder B. The retraction of both cylinders is represented by the complements of the inputs. Input signal c is an overload signal, indicating overloading during the extension of cylinder A. Outputs A_1 and B_1 should cause the extension of cylinders A and B, respectively, while their retraction is triggered by outputs A_2 and B_2. The network should perform the following task:
  a Cylinder A cycles once (extend and retract).
  b Cylinder B extends.
  c Cylinder A cycles again.
  d Cylinder B retracts, and the cycle repeats.
  e Cylinder A can be overloaded during its extension stroke, which must result in the retraction of cylinder A.
  f Retraction of cylinder A should relieve the overload condition of cylinder A; but if it does not, the circuit should return to its original state (by retraction of both cylinders).

By means of the State Matrix, Primitive Flow Table, Logic Specification Chart, and the Synthesis Table construct the logic description of the problem.

3 A door-actuating circuit has two inputs, a and b, and one output, Z. The inputs are generated by a pair of fluid push buttons and cannot be changed simultaneously. The output actuates the opening of the door, while its absence will result in the closing of the door automatically. The sequence of actuation of the inputs for opening the door is as follows:
  a Energize input a and maintain this energized condition.
  b Energize and deenergize input b twice.
  c Deenergize input a.
  d Energize input b and an output results (door opens).
  e The output Z will remain energized until input a is energized (door closes).
  f The circuit must be protected against all illegal sequences.

Using the LSC and the Synthesis Table construct the logic description of the problem.

4 A control network possesses two inputs, A and B, and two outputs, P and Q. The inputs change in a random sequence; however, they never change simultaneously. Output P is ON whenever A and B are both ON or when they are both OFF. Output Q occurs only when A is ON and B is OFF, *and* when output P has been activated and deactivated at least twice following the last deactivation of output Q. Use the LSC to construct the logic description.

5 A three-cylinder system consists of three equivalent cylinders, P, Q, and R, and three input signals, p, q, and r. The input-output circuit is constructed such that one input can represent the condition of each cylinder. Activation of the input indicates a fully extended condition, while a fully retracted position is indicated by the complement of the particular input. The input will remain in the condition arrived last until its complement occurs. Input p is used to represent the condition of cylinder P, q for cylinder Q, while r is used to represent the condition of cylinder R. The system is supposed to perform the following sequence:

$$P,\ \bar{P},\ Q,\ \bar{Q},\ Q,\ R,\ \bar{R},\ \bar{Q}\ Q,\ P,\ \bar{P},\ \bar{Q}$$

The symbols P and $\bar{P}$ are used to represent the extension and the retraction operation of cylinder P respectively. Use all descriptive techniques that are appropriate to represent this problem.

6 The three-cylinder system discussed in Problem 5 should perform the following sequence:

$$P,\ \bar{P}R,\ Q\bar{R},\ P\bar{Q},\ R,\ \bar{R},\ Q,\ \bar{P}Q,\ Q,\ R,\ P,\ \bar{P}\bar{Q}R$$

where the notation $P\bar{R}$ indicates the simultaneous extension and retraction of cylinders P and R respectively. Note that although the initiation of two or three cylinders is made simultaneously, the intended result may not occur at the same time. Use the most appropriate method for representing the problem.

7 A machine used for filling containers has two inputs and three outputs—one input, A, indicates that the container is in the proper position, while the other input, B indicates that the container is full. The three outputs, P, Q, and R, govern three power cylinders that are actuated by means of spring-offset,

LOGIC SYSTEM DESCRIPTION

four-way valves. The first cylinder, which is activated by the output P, is to move the container in place. The second cylinder, activated by output Q, pushes the container from the filling station to a nearby conveyor. Output R opens the valve that permits the filling of the container. The specification of the network has been devised as follows:

a Activate P until input A is activated, indicating the container is in position.
b Activate R until input B is ON (container is full).
c Activate Q until A has been deactivated for three seconds.
d After another three seconds, return to Step a.

Devise the logic description of the problem by using the most appropriate method. Hint: three-second delays are achieved by means of a TIME DELAY element. Generate a new output that activates this element and utilize the output of this delay element as a new input to the network.

8 A dump truck is to be designed such that it cannot lift its bed without first being shifted to a "PARK" position, and this action must also be preceded by the automatic opening of the rear end door. The specification of the logic network and its input and output circuits have been determined; it consists of five inputs, four outputs, and a delay element. The specification is as follows:

a If input B is activated (B is a manual signal to open the rear door), activate P_1 (output signal to open the door).
b If input C is ON (C is a manual signal to close the rear door), activate output P_2 (for closing the door).
c If input D is ON and A is OFF (A indicates that the truck is in "PARK" position and D is the signal for lifting the bed), no output should be generated.
d If both inputs A and D are ON, do the following:
   (1) If B was activated after C (indicating the door is open), then activate output Q_1 (to lift the bed).
   (2) If C was activated after B (indicating the door is closed), then activate P_1, followed by the activation of Q_1 after a delay of two seconds.
e If signal E is "ON" (E is a manual signal for the lowering of the bed), perform the following:
   (1) If B was activated after C, then activate output Q_2.
   (2) If B was activated prior to C, activate both outputs Q_2 and P_2.

Formulate the description of the logic system.

9 Describe all the problems presented in Chapter 3 using the LSC.

DEFINITIONS

Deterministic circuit A logic circuit in which only predetermined inputs are allowed.
Don't care output An output which is not required to satisfy the system function.
Event One discrete stage in a sequence.
Machine event A describable stage in the operating sequence of a machine.
Machine state An explicit description of the machine event.
Nonlogic elements Elements which do not perform a logic function.
Predetermined input An input generated as a result of a fixed sequence of events.
Random input An input occurring in a totally unpredictable or probabilistic manner.

Stable state A machine state that does not change without a change in the inputs.
Stochastic circuit A logic circuit in which predetermined, random, or both types of inputs are permissible.
Unstable state A transitory condition of the machine during which a machine attempts to reach an intended machine state.

REFERENCES

Cole, J. H., "Synthesis of Optimum Complex Fluid Logic Sequential Circuits," Ph.D. Dissertation, Oklahoma State University, Stillwater, Oklahoma, 1968.

Fitch, E. C., Jr., "Fluid Logic," Teaching Manual Published by the School of Mechanical and Aerospace Engineering, Oklahoma State University, Stillwater, Oklahoma, 1966.

Givone, D. C., *Introduction to Switching Circuit Theory*. New York: McGraw-Hill Book Co., 1970.

Huffman, D. A., "The Synthesis of Sequential Switching Circuits," *Journal of Franklin Institute*, Vol. 257, No. 3, pp. 161–190. March 1954.

Marcus, M. P., *Switching Circuits for Engineers*. Englewood Cliffs, N.J.: Prentice-Hall, Inc., 1962.

Maroney, G. E., "A Synthesis Technique for Asynchronous Digital Control Networks," M. S. Thesis, Oklahoma State University, Stillwater, Oklahoma, 1969.

Surjaatmadja, J. B., and Fitch, E. C. "Logic Specifications—Their Descriptions and Simplifications," Fluid Power Annual Research Conference, Paper No. P-75-56, Fluid Power Research Center, Stillwater, Oklahoma, October, 1975.

Woods, R. L., "The State Matrix Method for the Synthesis of Digital Logic Systems," M. S. Thesis, Oklahoma State University, Stillwater, Oklahoma, 1970.

Chapter 6

Combinational Logic Design

6-1 DESIGN METHODOLOGY

Design is the process of selecting and connecting elements to form a system. For a logic system, it is the process of finding a network that satisfies a prescribed set of requirements. Since Boolean algebra serves as the basis for logic design, applicable systems are limited to those having operations which can be described in binary terms.

The design of a logic system must begin with the selection and/or identification of the inputs and with a full description of the output requirements. Once the inputs and outputs are defined, a formal word statement of the problem can be developed. Based upon a knowledge of the inputs and outputs and the conditions under which the outputs must react with various inputs, a logic specification can be formulated. Such specifications were the subject of Chapter 5.

When dealing with combinational logic systems, a truth table for representing the implied logic is all that is needed, because it provides a unique accounting of the outputs for various combinations of the inputs. The output equations reflected by truth tables represent accurate logic descriptions of the required system. By using a simplification technique on these output equations (e.g., the Karnaugh map), the conventional design solution results. This solution involves a separate Boolean equation for each output and shows the interconnection of the various input variables.

Although the conventional solution of a combinational logic system is in its simplest normal form, there are a number of minimization techniques which can be applied to achieve "near optimal" form. These techniques take advantage of unspecified conditions in the logic specification, special features of the normal solution, and the unique characteristics of specific hardware and configurations of networks to establish the most desirable equations. Although the conventional solution may serve as the reference form initially, as various network minimization techniques are applied, the reference form is upgraded such that it represents the best solution for the network at any given time.

This chapter introduces various minimization techniques for obtaining practical network equations for combinational-type systems. Special consideration is given to equations and implementation requirements needed to obtain fast-response systems. The subject of network hazards in combinational systems is addressed, which represents a critical facet in network design. Finally, the computer techniques available for performing the various tasks needed in combinational logic design are introduced to facilitate the application of the methodology.

6-2 CONVENTIONAL SOLUTION

The conventional solution results from the use of the most elementary procedure for combinational networks. In this case, truth tables can effectively serve as the means for presenting the logic specification of the problem—a formal record of all the input-output relationships of the system. The truth table for design purposes is a minimal table in that only those combinations of inputs which have output significance are listed. Once this minimal truth table has been constructed, deriving the normal output equations for the network is very straightforward. Since these output equations are in "mixed logic" form, implementation of the normal solution is assumed to be by the fundamental elements AND, OR, and NOT.

The simplified version of the minimal truth table equations should contain the least number of literals and only prime implicants or implicates. It should be recognized that the minimum cost network of this type will result from the use of output equations which are either a sum of prime implicants or a product of prime implicates. The Karnaugh map as well as computerized tabular simplification methods are ideally suited for generating these all-important "prime" terms.

In order to formalize the design procedure for deriving the normal form of the output equations of combinational networks, the following essential steps are presented:

1 Construct the minimal truth table from the word statement.
2 Derive the output equations which are reflected by the minimal truth table.
3 Simplify the output equations by assuming that what is not specified in the minimal truth table are "zeros".
4 Implement the normal equations by the use of the fundamental operators (AND, OR, and NOT).

COMBINATIONAL LOGIC DESIGN

In order to illustrate the development of the normal combinational network, Example 6-1 is presented.

Example 6-1 A logic system has four inputs (a, b, c, and d) and two outputs (Z_1 and Z_2). Whenever the input combination bcd is 000, 111, or 100, output Z_1 must be activated while output Z_2 remains in its OFF condition. On the other hand, output Z_1 is deactivated and output Z_2 is activated whenever the input combination $bcd = 001$ exists. Furthermore, whenever bcd is equal to 010, both outputs must be activated. Signal a is the overriding inhibiting input—whenever it is activated, all outputs must be deactivated. The problem is to design the logic network.

The minimal truth table is shown in Table 6-1. Note that the input state at row 6 contains indeterminate entries and that not all input combinations have been listed. The output equations can be written directly from the minimal truth table as follows:

$$Z_1 = a'b'c'd' + a'bcd + a'bc'd' + a'b'cd' \tag{6-1}$$
$$Z_2 = a'b'c'd + a'b'cd' \tag{6-2}$$

By use of the Karnaugh map shown in Figure 6-1, the normal output equations for the network can be derived and are as follows:

$$Z_1 = a'c'd' + a'bcd + a'b'd' \tag{6-3}$$
$$Z_2 = a'b'c'd + a'b'cd' \tag{6-4}$$

The normal output equations can be implemented with the fundamental operators as illustrated in Figure 6-2. Note that only two input elements were employed and that INHIBITOR (AND-NOT, or passive NOT) elements were used because such two-input devices are commonly found in moving-parts hardware. The use of INHIBITOR elements also can eliminate the need of some individual inverters for inputs b, c, and d, since it is assumed that the complemented input signals are not directly available and have to be generated by the control network.

Implementing fluid logic circuits with multi-input elements can lead to network reduction in some cases. The two-input element circuit in Figure 6-2 can be

Table 6-1 Truth Table for Example 6-1

NO.	INPUTS				OUTPUTS	
	a	b	c	d	Z_1	Z_2
1	0	0	0	0	1	0
2	0	1	1	1	1	0
3	0	1	0	0	1	0
4	0	0	0	1	0	1
5	0	0	1	0	1	1
6	1	—	—	—	0	0

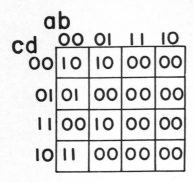

Figure 6-1 Karnaugh map for outputs Z_1 and Z_2.

compared with the multi-input element circuit in Figure 6-3. Certainly some reduction in the number of elements has been achieved; however, the scarcity of multi-input elements on the market limits their application in fluid logic systems.

6-3 UTILIZATION OF "DON'T CARES"

The significance of "don't care" terms in the simplification of Boolean equations was demonstrated in Chapter 2. In the actual design and synthesis of logic networks, the importance of "don't cares" cannot be overstressed. The reason for this is that in many cases, the requirements for a logic control system are not critical with respect to the outputs for a specific set of inputs. Such flexibility allows the designer to assume that those input states will never occur or that the system is actually unaffected by a given output resulting from a particular input state. For example, consider the case where a two-hand push button control system is to control the extension and the retraction of a high-powered cylinder. The push buttons, A and B, are to be pushed simultaneously by the operator in order for the cylinder to extend. Two other push buttons, C and D, also have to be actuated simultaneously to initiate the retracting operation of the cylinder. In designing the network for this system, it is quite logical to assume that the case where three or

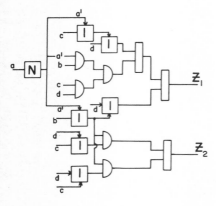

Figure 6-2 Two-input element implementation.

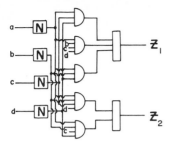

Figure 6-3 Multi-input element implementation.

four push buttons are actuated at the same time never exists, and therefore, "don't cares" can be assigned to these conditions.

Although all "don't care" terms must be designated as "zeros" or "ones" before the design is finished, the values should not be determined indiscriminately by the designer. By assigning the values of the "don't care" terms appropriately, a reduction in circuit complexity can be obtained.

The use of "don't care" term assignment for network reduction can be illustrated by considering the problem of Example 6-1. From the truth table of Table 6-1, it is evident that the input states $abcd$ = 0101, 0110, and 0011 were not assigned any output states. Assuming that there were no errors involved in the description of the problem, these terms are available for assignment as needed to reduce the network equations. The Karnaugh map in Figure 6-4 shows the location of these available terms. Instead of arbitrarily assigning zeros to the "don't cares" (as was the case for the normal solution), values are assigned optimally; hence, the prime implicants needed to represent the desired outputs become

$$Z_1 = a'b + a'd' = a'c + a'd' \tag{6-5}$$
$$Z_2 = a'b'c + a'b'd = a'c'd + a'cd' \tag{6-6}$$

The implementation of the output equations using two-input elements is illustrated in Figure 6-5. A 44 percent reduction in the number of elements was obtained over the normal solution implemented in Figure 6-2.

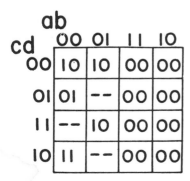

Figure 6-4 Karnaugh map with "don't care" terms for outputs Z_1 and Z_2.

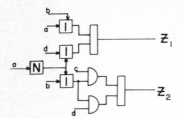

Figure 6-5 Network configuration using "don't cares" and two-input logic elements.

As can be appreciated from Example 6-1, "don't care" terms are not usually available from the problem statement. When the minimal truth table is reflected on a Karnaugh map, the "don't care" terms are vacant cells (not specified as zeros or ones). For small systems containing only a few inputs, the identification of the "don't care" terms represents no serious problem. However, for large-scale combinational networks, the generation of the "don't cares" can be a formidable task. In order to facilitate the identification of the "don't cares" contained in large-scale problem statements, a computer program was developed called NOR (a description of this program is given at the end of this chapter).

The simplification of large-scale networks by the use of strategic "don't care" term assignments can be illustrated by the problem depicted in the minimal truth table in Table 6-2. This ten-input, four-output control system identifies "don't cares" in the outputs for Z_1 (at row 4) and Z_4 (at rows 3, 4, and 7). Considering only the equation for output Z_1, the simplest normal equation (using the single "don't care" supplied) is:

$$Z_1 = abcfgh'i'j + a'b'c'd'e'gh'j + abcdeij' \qquad (6\text{-}7)$$

This equation does not consider the 2^{10} less 7 input states unspecified as to output conditions. Utilizing the cited computer program (NOR) to generate the "don't care" terms and the resulting prime implicants yields an output equation for Z_1 of

$$Z_1 = bf' + c'j + e \qquad (6\text{-}8)$$

and clearly shows a significant reduction in the number of literals needed to represent the required output.

Table 6-2 Truth Table for Ten-input Problem

No.	State	a	b	c	d	e	f	g	h	i	j	Z_1	Z_2	Z_3	Z_4
1	I_1	1	1	1	1	–	–	–	–	1	0	1	0	1	1
2	I_2	1	1	1	–	–	1	1	0	0	1	1	1	0	1
3	I_3	1	1	1	0	0	0	0	–	–	–	0	1	0	–
4	I_4	0	–	–	–	–	0	0	1	0	–	–	1	1	–
5	I_5	–	0	–	–	0	0	0	–	1	0	0	0	1	1
6	I_6	0	0	0	0	0	–	1	0	–	1	1	1	1	0
7	I_7	0	0	0	0	0	–	1	–	0	0	0	1	0	–

6-4 REDUCTION BY COMPLEMENTATION

Although it may be difficult to improve the reference form of the output equations resulting from the use of the "don't care" terms, consideration of the unique features of specific hardware may prove rewarding. In this regard, an awareness of the type of signals available for implementing a given network can play an important role in network element reduction. For example, unless complementing input signals are available (a rare occurrence), an output equation requiring complemented inputs can severely increase the complexity of the final network. Some of the factors which can help the designer optimize the network equations are

1. Know the type of elements that are available for use.
2. Know the active elements that can be used as passive elements.
3. Know the elements that have multiple inputs.
4. Know the type of available signals (complemented as well as uncomplemented types).
5. Know the form of the complemented functions for the available elements.

Obviously, it would be beneficial if the designer could recognize and subsequently select the logic elements best suited for the implementation of a given set of output equations. As this is not always possible, the transformation of the reference equations into equivalent (more recognizable) forms may enable significant reduction possibilities to be explored in order to obtain a more simplified solution.

The use of complementation of network equations offers a means of developing completely different but equivalent formats of the output equations. Hopefully, the implementation of the complement of the output equation can yield a simpler network. In order to illustrate the use of complementation as a means of reducing the complexity of networks, consider the problem explored in Example 6-1 and further improved by including the "don't cares." In this case, the Karnaugh map presented in Figure 6-4 is divided into two maps, one for each output, as illustrated in Figure 6-6. The portions of the maps enclosed by dashes represent the solutions

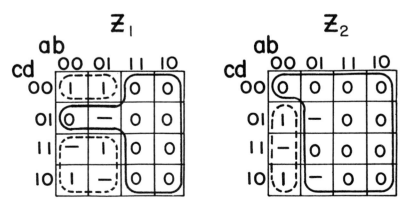

Figure 6-6 Karnaugh maps for Example 6-1.

expressed by Equations 6-5 and 6-6. The complementary functions of Z_1 and Z_2 can be obtained by considering the zeros and the "don't care" of the map and can be written as follows:

$$Z'_1 = a + c'd = a + b'd \qquad (6\text{-}9)$$
$$Z'_2 = a + b + c'd' \qquad (6\text{-}10)$$

The implementation of these equations using two-input elements and inhibitors results in the network schematically illustrated in Figure 6-7. In this particular case, no reduction of the network was obtained over that derived in Figure 6-5.

Still another form of the network equations can be obtained by complementing the output equations (Equations 6-9 and 6-10). This can be accomplished by the use of DeMorgan's theorem without increasing the number of AND's and OR's of the equations. The complementation of the above equations yields:

$$Z''_1 = Z_1 = a'(c + d') = a'(b + d') \qquad (6\text{-}11)$$
$$Z''_2 = Z_2 = a'b'(c + d) \qquad (6\text{-}12)$$

Implementing these equations with two-input elements including INHIBITOR's yields the network structure shown in Figure 6-8. Note that this time a substantial reduction in network complexity results.

When both INHIBITOR's and NOR's are available to the designer, the use of complemented solutions can be quite attractive for network reduction. For example, Figure 6-9 illustrates the implementation of Equations 6-11 and 6-12 using both INHIBITOR's and NOR's.

The reduction in network simplicity is even more striking. Note that in this situation it is important to recognize the inverted forms of these elements. For example, it is possible to rearrange Equation 6-11 and 6-12 to obtain the following:

$$Z_1 = a'(c + d') = a' \cdot \overline{c'd} = \overline{a + c'd} \qquad (6\text{-}13)$$
$$Z_2 = a'b'(c + d) = \overline{(a + b)}\,(c + d) \qquad (6\text{-}14)$$

The use of the complementation schemes advanced in this section were performed in order to minimize the number of individually complemented literals. An important facet of the complementation process is the employment of the AND-NOT and

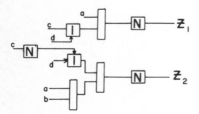

Figure 6-7 Results of complementation.

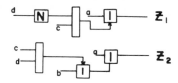

Figure 6-8 Results of double complementation.

the AND-NOR construction which uniquely avoids the need for using individual complementation of each literal.

6-5 MINIMIZATION BY FACTORING

Any technique permitted by the algebra which can transform the output equations into nonnormal form may enhance the implementation process. The use of factoring provides special versions of the equations which will enable novel solutions for network configurations. Basically, when an equation is factored, literals common to the factored terms are reduced in number. As an example of factoring, consider Equations 6-5 and 6-6. By factoring, one literal can be eliminated from Equation 6-5 and two literals from Equation 6-6, as shown below:

$$Z_1 = a'(c + d') \qquad (6\text{-}15)$$
$$Z_2 = a'b'(c + d) \qquad (6\text{-}16)$$

In this particular case, these equations are identical to those derived from complementation (Equations 6-11 and 6-12) and the implementation would be the same as that shown in Figure 6-8. The reader should note, however, that this is an unusual case and normally factoring yields a different equation form than complementation.

As straightforward as it may seem, optimal factoring may pose an extremely difficult task for the designer. Trial and error and also the intuition of the designer may be an important factor for achieving the optimal solution. The distributive law plays an important role in the factoring approach, which is basically:

$$XY + XZ = X(Y + Z) \qquad (6\text{-}17)$$

where X, Y, and Z may be any Boolean functions. No particular algorithm exists for obtaining optimal factoring; however, the following set of rules may serve this purpose:

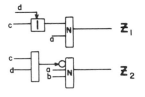

Figure 6-9 Implementation of Example 6-1, using OR, NOR, and INHIBITOR elements.

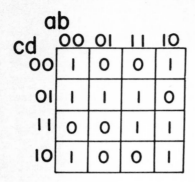

Figure 6-10 Karnaugh map for Example 6-2.

1. Observe whether there are literals that occur in all terms of the group under consideration. Factor these literals out of the group.
2. Select a literal that occurs most in the representation. Divide the group into two groups—one containing the particular selected literal and the other that does not contain the literal.
3. Perform Steps 1 and 2 iteratively for each established group until no other factoring can take place. In most cases, this process will result in a near minimal factoring when two-input elements are used.

It should be realized that when multiple fan-in elements are used, factoring may not be desirable (partial factoring may be all that is needed to achieve an optimal form). As a comprehensive illustration of factoring, consider Example 6-2.

Example 6-2 A logic system has four inputs and one output, and can be described by the following:

$$Z = a'b'c'd' + ab'c'd' + a'b'c'd + a'bc'd + abc'd + abcd + ab'cd \\ + a'b'cd' + ab'cd' \qquad (6\text{-}18)$$

No "don't care" conditions are available.

The Karnaugh map for this problem can be constructed as shown in Figure 6-10. This map indicates the simplified form of the equation is:

$$Z = b'd' + a'c'd + abd + acd \qquad (6\text{-}19)$$

Since signal d occurs most frequently, grouping per Step 2 is performed:

$$Z = b'd' + (a'c'd + abd + acd) \qquad (6\text{-}20)$$

and by Step 1 d is factored out of the group:

$$Z = b'd' + d(a'c' + ab + ac) \qquad (6\text{-}21)$$

COMBINATIONAL LOGIC DESIGN

Similarly, grouping can be performed within the group on the right:

$$Z = b'd' + d(a'c' + a(b + c)) \tag{6-22}$$

This last equation completes the factoring process of Equation 6-19.

It is also important to realize that a combination of complementing and factoring is a powerful means of transforming equations into other equivalent forms. For example, from the Karnaugh map in Figure 6-10 the complementary equation for the output is:

$$Z' = bd' + a'cd + ab'c'd \tag{6-23}$$

and factoring yields:

$$Z' = bd' + d(a'c + ab'c') \tag{6-24}$$

Taking the complement of Equation 6-24 gives:

$$Z = (b' + d)(d' + (a + c')(a' + b + c)) \tag{6-25}$$

Equation 6-25 can be obtained in another way by applying DeMorgan's rule upon Equation 6-24 to obtain:

$$Z = (b' + d)(a + c' + d')(a' + b + c + d') \tag{6-26}$$

and then factoring.

It should be obvious by now that the selection of the network equation form is very important to the implementation process. The one that will actually produce the minimal network depends upon the available logic elements. For example, if multi-input elements are used, the selection of Equation 6-26 is desired since only eight elements are necessary to implement the network (including individual inverters for input signals a, b, c, and d). The completed network diagram can be seen in Figure 6-11. On the other hand, if two-input elements are used and inhibitors are

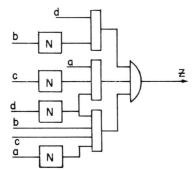

Figure 6-11 Network configuration for Example 6-2, using universal multi-input logic elements.

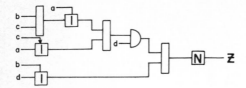

Figure 6-12 Network configuration for Example 6-2, using two-input elements.

available, the selection of Equations 6-22, 6-25, and 6-26 produces the minimal solution. During the implementation, factoring may again be used to avoid any individual complementing of the input signals and taking advantage of the features of the INHIBITOR element. For example, Equation 6-25 can be reformulated as follows:

$$Z = \overline{(bd')} \, \overline{(d' + \overline{(a'c)} \, \overline{(a(\overline{b+c}))}} \tag{6-27}$$
$$Z = (bd') + \overline{(d(a'c + a\overline{(b+c)}))} \tag{6-28}$$

The unique representation of Equation 6-28 may be observed in Figure 6-12.

Additional reduction in the network complexity can also be achieved if, for example, all types of elements are available, including the NOR and COINCIDENCE functions. Implementing Equation 6-21 with these elements produces the minimal form of the network, as shown in Figure 6-13. Note that Equation 6-21 was actually not an intended solution but was an intermediate function generated during the process. Example 6-2 vividly illustrates the actual complexity involved when a large variety of logic elements are available for the implementation of a network. The reader may also be aware of still other possibilities which were not brought to the limelight in this example. Generating all the variations may pose an endless task especially for the extremely large network functions. The practicality of the minimization effort depends upon the cost-saving in hardware and labor.

6-6 MULTITERMINAL NETWORKS

In previous sections of this chapter, the synthesis of the network was performed on the basis that only one output existed in the system. In cases where more than one output was present, each output was treated or synthesized individually. This was done without regard to the influences each output may have on other outputs when considering the network as a whole. Such disregard for complementary interaction of outputs eliminates many possibilities for the multiple use of generated signals.

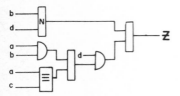

Figure 6-13 Network for Example 6-2, including NOR and COINCIDENCE.

COMBINATIONAL LOGIC DESIGN

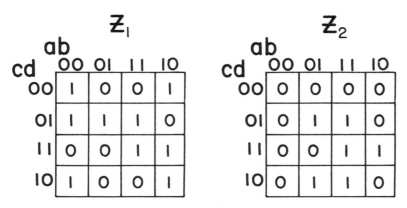

Figure 6-14 Karnaugh maps for Example 6-3.

The concept of multiterminal networks (those exhibiting more than one output) and simplifying the network as a whole can be illustrated using Example 6-3.

Example 6-3 A two-output logic system is described by the Karnaugh maps shown in Figure 6-14. By synthesizing each output individually, the following equations can be obtained:

$$Z_1 = b'd' + a'c'd + abd + ab'c \qquad (6\text{-}29)$$
$$Z_2 = bc'd + acd + bcd' \qquad (6\text{-}30)$$

Implementation of these equations yields the fifteen-element network shown in Figure 6-15. Applying both factoring and complementation techniques to the above output equations results in a somewhat simpler set of equations:

$$Z_1 = \overline{(bd')}\,\overline{(a'cd) + (ad(b+c))} \qquad (6\text{-}31)$$
$$Z_2 = bc'd + acd + bcd' \qquad (6\text{-}32)$$

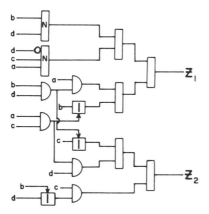

Figure 6-15 Network for Example 6-3 obtained by individual output synthesis.

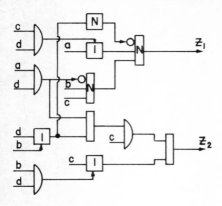

Figure 6-16 Network for Example 6-3 obtained by individual factoring and complementing.

These minimal network equations require 12 two-input elements to satisfy the network, as can be seen in Figure 6-16. It can be noted that the implementation of these equations demonstrates the importance of "reusing" generated signals and synthesizing the total system as a whole.

From the Karnaugh maps of both outputs (Figure 6-14), it can be noted that the cells $abcd = -101$ and $1-11$ occur in both representations of Z_1 and Z_2. By including these two terms in the representation of the network, the output equations become:

$$Z_1 = b'd' + a'c'd + bc'd + acd \qquad (6\text{-}33)$$
$$Z_2 = bc'd + acd + bcd' \qquad (6\text{-}34)$$

and they may be implemented by the eleven-element network shown in Figure 6-17. Note that factoring was performed such that the common factor between the two functions was used in both outputs.

In Example 6-3, the common factors for outputs Z_1 and Z_2 happened to be the prime implicants of both the output equations. In this particular case, perhaps simple intuition may suffice for the selection of a near minimal network; however, this is not generally the case. To further illustrate the use of multiterminal networks, consider Example 6-4.

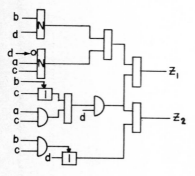

Figure 6-17 Network for Example 6-3 obtained by simultaneous synthesizing of the outputs.

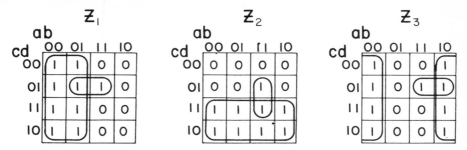

Figure 6-18 Karnaugh map for Example 6-4.

Example 6-4 A logic system has three individual outputs and four input variables and can be described by the Karnaugh map shown in Figure 6-18. Individual synthesis of each of the outputs results in the following minimal equations:

$$Z_1 = a' + bc'd \tag{6-35}$$
$$Z_2 = c + abd \tag{6-36}$$
$$Z_3 = b' + ac'd \tag{6-37}$$

Based upon these equations, a ten-element network is needed to implement the network and this schematic is shown in Figure 6-19.

When the output circuit equations are derived considering multiple-output implementation, the following equations result:

$$Z_1 = a' + abc'd \tag{6-38}$$
$$Z_2 = c + abc'd \tag{6-39}$$
$$Z_3 = b' + abc'd \tag{6-40}$$

To implement these output equations requires the eight-element network given in Figure 6-20. This example again reveals the importance of considering the output requirements as a unit instead of synthesizing them individually.

It is important to realize that the process of design for multiple-terminal networks reverts to simply identifying cells or subcubes in the Karnaugh maps

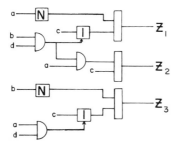

Figure 6-19 Circuit diagram for Example 6-4 obtained by individual output synthesis.

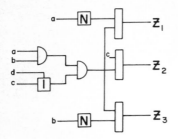

Figure 6-20 Circuit diagram for Example 6-4 obtained by multiple output synthesis.

which are common to several of the output maps. In order to reveal the characteristic features of the method, consider Example 6-5.

Example 6-5 A four-input, three-output logic system is described by the Karnaugh maps shown in Figure 6-21. It can be noted that cells $abcd = 1000$ and 0101 are common to all three outputs. Therefore, these terms should be generated only once and used appropriately by all outputs of the system. When this is accomplished, the equations can be written as follows:

$$Z_1 = c + c'\,(a'bd + ab'd') \tag{6-41}$$
$$Z_2 = cd' + ab + c'\,(a'bd + ab'd') \tag{6-42}$$
$$Z_3 = cd + a'b' + c'\,(a'bd + ab'd') \tag{6-43}$$

These output equations can be implemented using the fourteen-element network shown in Figure 6-22.

If this system was synthesized without consideration of the multiple-terminal aspects, the following output equations in near-minimal form would result:

$$Z_1 = c + a'bd + ab'd' \tag{6-44}$$
$$Z_2 = cd' + ad' + ab + bc'd = d'\,(c + a) + b(a + c'd) \tag{6-45}$$
$$Z_3 = a'b' + cd + a'd + b'c'd' = b'\,\overline{(a(c + d))} + d\,\overline{(ac')} \tag{6-46}$$

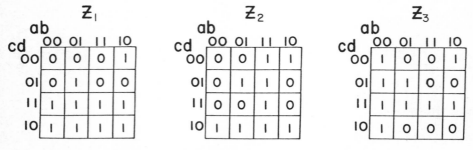

Figure 6-21 Karnaugh map for Example 6-5.

COMBINATIONAL LOGIC DESIGN

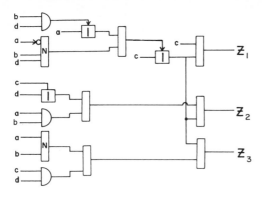

Figure 6-22 Network realization of Example 6-5 by multiterminal synthesis.

The network needed to satisfy the output equations (Equations 6-44, 6-45, and 6-46) would require seventeen elements, as illustrated in Figure 6-23. It should be realized that the multioutput synthesis procedure does not guarantee a minimal network, but it does give another possibility for reducing the network representation.

The process of generating the multiple-output prime implicants and implicates for large-scale problems becomes an almost impossible task for the designer when intuitive means are applied. To facilitate this important minimization process, a computer program (called subroutine MOMIN) has been developed. The description and use of this program is discussed in the last section of this chapter.

6-7 UTILIZATION OF LOGICALLY COMPLETE ELEMENTS

No discussion on combinational logic network reduction would be complete without considering the unique features offered by logically complete elements—specifically, the NOR, the NAND, and the INHIBITOR. An element is said to be logically complete when it can be used to represent any logic function. Thus, it is possible to construct logic networks using only one type of logically complete element.

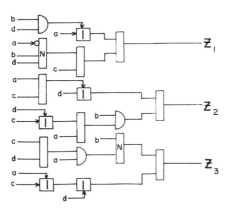

Figure 6-23 Network realization of Example 6-5 by individual output synthesis.

Figure 6-24 Obtaining the NOT function by a NOR element.

Since a logic function can be completely described by the operators AND, OR, and NOT, it is important to show that NOR, NAND, and INHIBITOR logic elements can be utilized to represent any of the above basic logic operators. In order to demonstrate this feature, consider the following function of an n-input NOR element:

$$f_{nor} = \overline{a_1 + a_2 + a_3 \ldots \ldots + a_n} \tag{6-47}$$

For a NOT function, $a_2, a_3, \ldots a_n$ are set equal to zero (exhausted), which results in:

$$f_{nor} = \overline{a_1 + 0 + 0 + \ldots \ldots + 0} = a_1' \tag{6-48}$$

The symbolic diagram for this operation can be observed in Figure 6-24.

To represent an AND operation, Equation 6-47 can be transformed as follows:

$$f_{nor} = a_1' \cdot a_2' \cdot a_3' \cdot \ldots \ldots \ldots \cdot a_n' \tag{6-49}$$

and for the case where $a_i = b_i'$ and $1 \leq i \leq n$, then:

$$f_{nor} = b_1 \cdot b_2 \cdot b_3 \cdot \ldots \ldots \cdot b_n \tag{6-50}$$

which is clearly a representation for an AND operation, as illustrated in Figure 6-25.

The OR-operation can be devedloped from Equation 6-47 in a manner similar to that performed for the AND as follows:

Let $a_1 = f_{nor2} = \overline{b_1 + b_2 + \ldots \ldots + b_n}$ while $a_2, a_3, \ldots a_n$ are set equal to zero (exhausted). Substituting appropriately into Equation 6-47 yields:

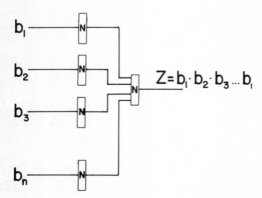

Figure 6-25 AND function developed by NOR elements.

COMBINATIONAL LOGIC DESIGN

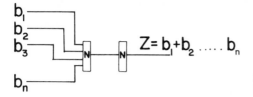

Figure 6-26 OR function developed by NOR elements

$$f_{nor} = \overline{\overline{b_1 + b_2 + b_3 + \ldots\ldots + b_n} + 0 + \ldots + 0}$$
$$= b_1 + b_2 + b_3 + \ldots\ldots + b_n \qquad (6\text{-}51)$$

This OR representation is symbolically depicted in Figure 6-26.

The methods used to represent the basic logic operators with NOR elements can be performed for the NAND and INHIBITOR logic devices just as easily. It should be apparent that since these contractive elements exhibit the remarkable feature of functional completeness, a logic circuit can be constructed solely by either NOR, NAND, or INHIBITOR elements. This feature gives the user the convenience of dealing with one single-type element and eliminates the need for maintaining a large inventory of various logic elements.

NOR logic devices have one other advantageous feature of importance. They are relatively simple to construct and are available in most types of logic elements—both gas and liquid. The incorporation of a multi-input feature poses no problems for the NOR; in fact, most NOR logic devices are basically multi-input units. This multi-input feature offers significant advantages—a faster, fewer-element, and more likely a less costly logic network.

The number of logic elements which must be actuated in order to generate a particular output has a direct influence on the speed of response of the output signal. Hence, by using logic elements which possess high fan-in and fan-out opportunities, the number of "levels" or "stages" a signal needs to traverse to reach the output can be reduced. The general concept of signal actuation levels or stages can be appreciated by the illustration given in Figure 6-27. Each level represents a

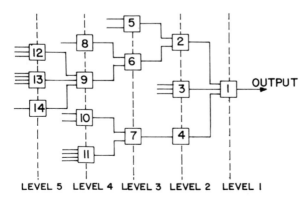

Figure 6-27 A generalized multilevel circuit.

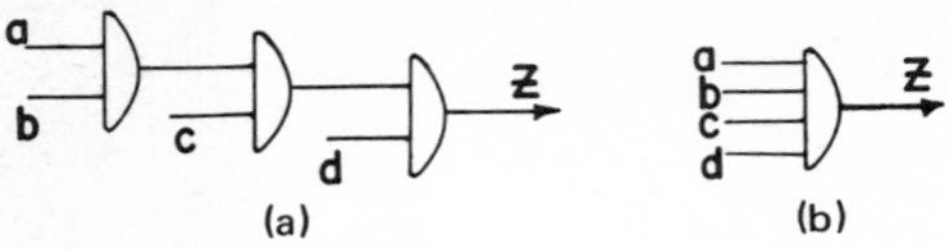

Figure 6-28 Circuits representing $Z = a \cdot b \cdot c \cdot d$: (*a*) Implemented by two-input AND values; (*b*) implemented by four-input AND values.

signal delay potential and should be considered in the design of a logic system. For example, in Figure 6-28, four inputs are controlled by two-input (fan-in) elements producing a three-level network, while the same four inputs can be controlled as illustrated with a multi-input element in a single-level network. As the difference in switching time between multi-input elements and two-input elements is usually insignificant, the network exhibiting the least number of levels is generally the fastest. Here lies one of the subtle advantages of the NOR logic element.

In Chapter 4, the rules for transforming normal Boolean equations into NOR and NAND form were presented. The rules of DeMorgan were applied in the following manner:

$$\begin{aligned} Z &= ab' + a'c + bc' = \\ &\overline{\overline{\overline{ab'}} + \overline{\overline{a'c}} + \overline{\overline{bc'}}} = \\ &\overline{\overline{(a' + b)} + \overline{(a + c')} + \overline{(b' + c)}} \end{aligned} \tag{6-52}$$

which constitutes a transformation to the NOR format. For the NAND form, the above equation is provided as:

$$\begin{aligned} Z &= ab' + a'c + bc' = \\ &\overline{\overline{ab' + a'c + bc'}} = \\ &\overline{\overline{a\cdot b'} \cdot \overline{a\cdot c'} \cdot \overline{b\cdot c'}} \end{aligned} \tag{6-53}$$

When the function is not normal, DeMorgan's rule is performed repetitively, as for example:

$$Z = a'b(c + d'e(f' + g'h)) = \tag{6-54}$$

$$\begin{aligned} &\overline{\overline{a'b(c + d'e(f' + g'h))}} = \\ &\overline{a + b' + \overline{(c + d'e(f' + g'h))}} = \\ &\overline{a + b' + \overline{(c + \overline{\overline{d'e(f' + g'h)}})}} = \\ &\overline{a + b' + \overline{c + \overline{d + e' + \overline{f' + \overline{g + h'}}}}} \end{aligned} \tag{6-55}$$

which is the NOR form. This transaction can be performed more directly by using the rules that are listed below:

COMBINATIONAL LOGIC DESIGN 157

1. Express the function in a product of sums of products ... format.
2. Determine and label the levels of each operator of the function. (See "Definitions" for the term level.)
3. Starting at the highest level, NOR operators are substituted for each group of operators which are not separated by a lower-level operator. The NOR operator can be a single operator type, $\overline{+}$, or of multiple form, $\overline{(+ + + \ldots +)}$.
4. Complement all individual literals that appear at odd numbered levels.

Note that the above rule also applies for the NAND transformation with a minor change, i.e., the function should be represented initially by a sum of product of sums of products of .... form, and that instead of NOR operators, NAND operators are used.

As an example of the use of the procedure, the levels in Equation 6-54 can be indicated as follows:

$$Z = a' \cdot b \cdot (c + d' \cdot e \cdot (f' + g \cdot h)) \qquad (6\text{-}56)$$
$$1\ 123345$$

The levels are indicated below each operator. The determination of these levels is made easier when the proper group of terms or literals are all enclosed in parentheses. Then, the innermost parentheses will always assume a higher level than the next outer parentheses. The above equation is already in its products of sum .... format, and hence no changes in the format are necessary. According to Rule 4, the literals a', b, d', e, g', and h should be complemented, as they appear at odd levels. Performing the replacement of all operators by NOR operators results in the formation of Equation 6-55.

In the above example, the format of the equation already satisfies the requirement of the algorithm (indicated in Step 1). However, when this is not satisfied, the multiplication (AND-ing) of the function with the unit term 1 or $0'$ can be utilized in solving the problem. For example, if:

$$Z = a' + b + c \cdot (d' + e + f' \cdot (g' + h)) \qquad (6\text{-}57)$$

it can be transformed into a product-of sums-of products-of ... form by AND-ing the function with a logic 1 (or $0'$) as follows:

$$Z = (a' + b + c \cdot (d' + e + f' \cdot (g' + h))) \cdot 0' \qquad (6\text{-}58)$$
$$22344561$$

In accordance with Step 4, complementation has to be performed upon the literals $0'$, c, and f' to obtain the NOR format:

$$Z = \overline{\overline{0 + a' + b + \overline{c' + d' + e + \overline{f + g' + h}}}} \qquad (6\text{-}59)$$

Similarly, when Equation 6-54 is to be transformed into a NAND form, the sum-products-of-sums ..... format can be obtained by adding (or OR-ing) the Boolean value 0 or $1'$ as follows:

$$Z = a' \cdot b \cdot (c + d' \cdot e \cdot (f' + g' \cdot h)) + 1' \qquad (6\text{-}60)$$
$$\;\; 2 \;\; 2 \;\;\;\; 3 \;\;\;\; 4 \;\; 4 \;\;\;\; 5 \;\;\; 6 \;\;\;\; 1$$

The NAND form would then be:

$$Z = \overline{\overline{\overline{1 \cdot a' \cdot b \cdot \overline{c' \cdot d' \cdot e \cdot \overline{f \cdot \overline{g' \cdot h}}}}}} \qquad (6\text{-}61)$$

In Equation 6-59, it can be shown that a six-element network using three input-NOR logic elements is suitable to represent the logic system. Unfortunately, this is only true when both conditions of the signals exist—the true and complementary signals. Otherwise, the complements must be generated by using separate NOR elements. Thus, in order to avoid the need of these extra NOR elements, an attempt should be made to place the complementary literals of the function in the odd-level position of the equation. Consider, for example, the following equation:

$$Z = a'b'c'd' + ac + bc + cd \qquad (6\text{-}62)$$

which, for the application of NOR elements, has to be transferred into the products-of-sum-of ...... format. Hence:

$$Z = (a'b'c'd' + ac + bc + cd) \cdot 1 \qquad (6\text{-}63)$$
$$\;\; 3\;3\;3 \;\;\;\; 2\;3 \;\;\; 2\;3 \;\;\; 2\;3 \;\;\;\; 1$$

Note that the literals a, b, c, and d are all at the odd-level position, which requires the complementation:

$$Z = \overline{\overline{\overline{a} + \overline{b} + \overline{c} + \overline{d} + a' + c' + b' + c' + c' + d'}} \qquad (6\text{-}64)$$

The resulting network can be observed in Figure 6-29. Note that individual complementation was used, since it is necessary to assume that no complementary literals are available in the system. Sometimes the transformation of an output equation into another form will change the location of the odd-level positions and allow a much simpler network to be constructed. For example, consider Equation 6-62 and perform a simple factoring operation as follows:

$$Z = a'b'c'd' + c \cdot (a + b + d) \qquad (6\text{-}65)$$

which, by the distributive law can be transformed into:

$$Z = (a' \cdot b' \cdot c' \cdot d' + c) \cdot (a' \cdot b' \cdot c' \cdot d' + a + b + d) \qquad (6\text{-}66)$$
$$\;\; 3 \;\;\; 3 \;\;\; 3 \;\;\; 2 \;\;\; 1 \;\;\;\;\; 3 \;\;\; 3 \;\;\; 3 \;\;\; 2 \;\; 2 \;\; 2$$

COMBINATIONAL LOGIC DESIGN

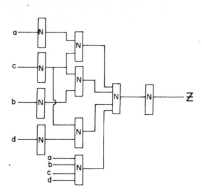

Figure 6-29 NOR network representation of Equation 6-62 by direct implementation.

and thus the NOR form is:

$$Z = \overline{\overline{a+b+c+d} + c + \overline{a+b+c+d} + a + b + d} \tag{6-67}$$

Hence, by the above manipulation, the complemented literals have been "moved" to odd-level positions; therefore, no individual complementation is needed in the construction of the network. The resulting four-element network is shown in Figure 6-30. A drastic reduction in elements was achieved when compared to the resulting network needed to implement Equation 6-64. It should be stated that it is not always possible to displace the complemented literals to odd-level locations.

INHIBITOR elements are advantageous only for particular circuits. For example, Equation 6-56 can be successfully transformed into an INHIBITOR-type function as follows:

$$Z = ((((\overline{\overline{(g' \cdot h)} \cdot f)} \cdot e) \cdot d') \cdot c') \cdot b) \cdot a' \tag{6-68}$$
$$7\phantom{\overline{\overline{(g'}}}654321$$

Note that in this particular example, the operator at level (3) deals with two complemented quantities and therefore an individual INHIBITOR element should be used for complementing one of the quantities. It is therefore quite obvious that if many such situations occur in the representation, the degree of complexity of the network will be highly increased, which contributes to the impracticality of the sole use of the INHIBITOR element.

6-8 THREE-LEVEL NOR LOGIC

In the previous section, the concept of actuation stages or levels was introduced. However, no limitation was imposed on the network regarding the maximum

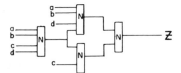

Figure 6-30 Results of "moving" complemented literals to odd levels.

number of levels to which a network should be restricted. It was emphasized that increasing the number of levels in a circuit tends to decrease the speed of response of a network.

In general, it is practical and feasible to reduce the number of actuation levels of a network to three when only uncomplemented inputs are available. The three levels are the following functions:

- Disjunctive function—perform the OR operation.
- Conjunctive function—perform the AND operation.
- Complementing function—perform the NOT operation.

In order to satisfy a network implementation criterion of allowing a maximum of three levels of actuation, almost unlimited fan-in characteristics must be exhibited by the network logic elements. Such characteristics are exhibited by NOR logic elements, as they are basically multi-fan-in elements. Moreover, for some available passive-type NOR elements, the fan-in number can be further "expanded" by the series connection shown in Figure 6-31. Note that the connection permits an uninhibited passage through the valves, and hence, only single-input changes are necessary to "switch" the output of the series. Therefore, the combination will have a switching speed comparable to that of a single valve, although it might be slightly impaired by the viscous effects and the capacitance of longer flow passages.

As a means of illustrating some of the possibilities offered by "three-level NOR logic," and to formalize rules for implementing these networks, consider Example 6-6.

Example 6-6 This problem deals with the implementation of the network represented by the disjunctive Karnaugh map shown in Figure 6-32 by the use of NOR logic elements. By representing a network in a normal product-of-sums form, the number of levels in a circuit will always be equal to or less than three. This is due to the fact that the necessary individual complementation of the input signals at the first level is always included. For this example, the normal product of sums can be written directly from the Karnaugh map as follows:

$$Z = (a + b + c')(a + b + d')(a + b' + c + d)(a' + b + c + d) \qquad (6\text{-}69)$$

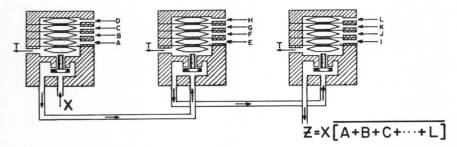

Figure 6-31 Series connection of NOR elements.

COMBINATIONAL LOGIC DESIGN

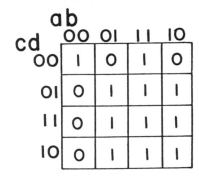

Figure 6-32 Karnaugh map for Example 6-6.

Implementing this equation with NOR elements results in the network shown in Figure 6-33. Observe that in general, the three levels of actuation represents the following signals:

- Level 1—output signal.
- Level 2—true valued variables.
- Level 3—complemented variables.

The task of designing a minimal network is often difficult, since the "best" form is generally obscure. However, there is good reason now to accept the fact that the minimum form actually results from achieving a minimal-level format. One method which can be used to establish this minimal-level network is based upon the following algebraic rules:

Rule 1 Within a sum term, any complemented literal can be replaced by a product of solely complemented literals, consisting of the complemented literal under consideration and the complement of one or more true-valued literals within the term.

Basically, this rule stems from the law of reflection, i.e.:

$$A + A'B' = A + B' \tag{6-70}$$

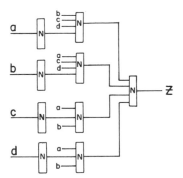

Figure 6-33 Implementation of Equation 6-69.

It can be illustrated by considering the following sum term:

$$T = (a' + b' + c + d + \ldots\ldots + z) \qquad (6\text{-}71)$$

Assuming that a and b are the only complemented literals in the sum term, the equation will not be altered by the following changes:

$$T = (a'c'd' + b'c'd' \ldots z' + c + d + \ldots z) \qquad (6\text{-}72)$$
$$= (a'c'd' \ldots p' + b'm'o' \ldots x' + c + d + \ldots z) \qquad (6\text{-}73)$$

Basically, this rule indicates that one or more inputs to an n-th level NOR logic element can also be represented at an $(n + 1)$th-level NOR element leading to the n-th level element under discussion, without altering the logic of the output. For example, consider the network shown in Figure 6-34(a). Representation of input a in the $(n + 1)$th level would not change the function of the output (see Figure 6-34(b). Similarly, the modification performed on the circuit shown in Figure 6-34(c) results in the identical circuit of Figure 6-34(d). It is, therefore, apparent that the establishment of Rule 1 accommodates virtually endless means for manipulating NOR logic networks.

The second rule is based upon the law of distribution.

Rule 2 Two or more sum terms that are connected conjunctively can share one second-level NOR element if they contain one or more literals in common and if each term contains at least one uncommon complemented literal that occurs only once in the expression. The input to this second-level NOR element consists of the common literals and the distributive multiplication of the uncommon, complemented literals. Each of these distributively multiplied literals can be generated by a third-level NOR element. By Rule 1, each of the products may be multiplied by one

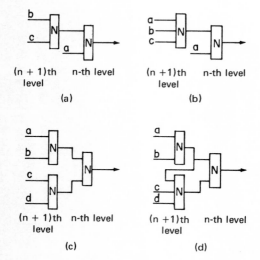

Figure 6-34 Equivalent transformations by Rule 1.

COMBINATIONAL LOGIC DESIGN

or more complements of the true-valued or uncomplemented common literals for simplification purposes.

Essentially, the rule states that the following expressions are equivalent:

$$T = (a + b')(a + c' + d')(a + e' + f')$$
$$= (a + b'c'e' + b'c'f' + b'd'e' + b'd'f') \tag{6-74}$$

which is quite evident from the standpoint of Boolean algebra. As a corollary, the following special condition is valid:

$$T = (a + b')(a + c')(a + d')$$
$$= (a + b'c'd') = (a + a'b'c'd') \tag{6-75}$$

The situation depicted by Equation 6-75 is most important for the minimization of NOR logic circuits. This is due to the fact that the application of Rule 2 upon such networks will always result in the reduction of the circuit components. In order to familiarize the reader with the procedure of manipulating NOR logic networks, a pictorial proof of this rule is given.

Consider, for example, the network shown in Figure 6-35(a), which represents $Z = (a + b')(a + c')$. According to Rule 1, the lower branch can be duplicated and placed as an input to NOR element No. 3 of Figure 6-35(b) (see elements 1 and 2). Because of the fact that signal a is represented at both elements 2 and 3, elimination of a at NOR element 2 would not affect the logic of the circuit, thus leaving a double complementation of signal c. Therefore, NOR elements 1 and 2 can be eliminated by applying signal c to element No. 3. By Rule 1, c can also be assigned to NOR element No. 4. Performing the above procedure backwards on signal c results in the reestablishment of elements 1 and 2, as shown in Figure 6-35(c). The elimination of these two elements follows due to the similarity to the lower branch shown.

The process can be performed again by duplicating the upper branch and assigning it to NOR element No. 3, as shown in Figure 6-35(d). Temporarily eliminating NOR elements 1 and 2 results in the equivalent circuit shown in Figure 6-35(e). Input signal b can hence be "drafted" and assigned to NOR element 4, following which the reinstallment and elimination of elements 1 and 2 can be performed. This leaves two identical branches feeding into the first-level NOR element; therefore, merging of the two branches is made, which results in the three-element network shown in Figure 6-35(f). Note that this simplification can also be obtained by performing the law of distribution as demonstrated by Equation 6-75.

Rule 3 Any group of n-prime implicates of an m-variable system can share a common NOR logic element if there are n_1 variables where for each prime term they are all represented by 1's with the exception of one single "0." The common NOR element will have as its input all the n_1 variables, and will use them to replace the complemented variables involved.

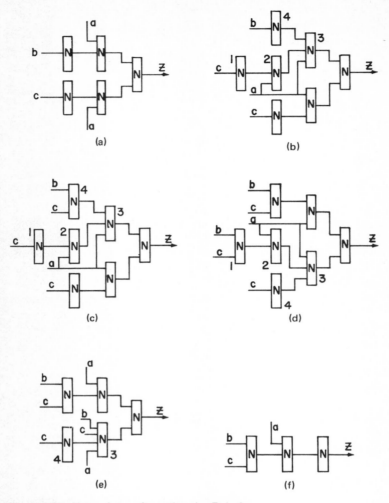

Figure 6-35 Network transformations by Rule 1.

Rule 3 is an extension of Rule 1 for accommodating a specific application in the three-level NOR logic implementation. As an example of this rule, consider the following conjunctive equation:

$$Z = (a + c' + d + e + f' + g + h)(a' + b + d + e' + f + g + h)$$
$$(a' + b + c + d + e + f' + g + h)(a + b + c + d' + e + f + g + h)$$
$$(a + b' + c' + d' + e + f + g + h) \qquad (6\text{-}76a)$$

This equation can be presented in matrix format as follows:

COMBINATIONAL LOGIC DESIGN

$$Z = \begin{bmatrix} a & b & c & d & e & f & g & h \\ 1 & - & 0 & 1 & 1 & 0 & 1 & 1 \\ 0 & 1 & - & 1 & 0 & 1 & 1 & 1 \\ 0 & 1 & 1 & 1 & 1 & 0 & 1 & 1 \\ 1 & 1 & 1 & 0 & 1 & 1 & 1 & 1 \\ 1 & 0 & 0 & 0 & 1 & 1 & 1 & 1 \end{bmatrix} \tag{6-76b}$$

The above equation can have variables d, e, f as one of its common factors. Therefore, these variables, fed into a single NOR element, can be utilized to represent the negated literals d, e, and f. Other common factors that can be found are a,b and a,c. The simplified function can therefore be represented by:

$$Z = (a + a'c' + d + e + d'e'f' + g + h)(a'b' + b + d + d'e'f' + f + g + h)(a'c' + b + c \ldots)\ldots \tag{6-77}$$

The reader is encouraged to prove this representation diagrammatically by means of Rule 1.

The three-level NOR logic implementation method is initiated by expressing the switching function in its most simplified conjunctive form. This can be obtained most conveniently from the Karnaugh map. The resulting conjunctive expression is then inspected for any possibilities where the conditions mentioned in the above rules occur.

As an example, consider the problem cited in Example 6-6. This time, Equation 6-69 is presented in matrix format to enhance the inspection procedure:

$$Z = \begin{bmatrix} a & b & c & d \\ 1 & 1 & 0 & - \\ 1 & 1 & - & 0 \\ 1 & 0 & 1 & 1 \\ 0 & 1 & 1 & 1 \end{bmatrix} \begin{matrix} (1) \\ (2) \\ (3) \\ (4) \end{matrix} \tag{6-78}$$

It can be observed that terms 1 and 2 satisfy Rule 2, while terms 3 and 4 satisfy Rule 3. Hence, the complemented variables c' and d' can become a common factor for terms 1 and 2, which become one sum-of-products expression. On the other hand, by Rule 3, a' and b' are selected to serve as a common factor of terms 3 and 4. Because all other variables of terms 3 and 4 are 1's, their common factor can be "expanded" to cover all variables—this is also true for the case of the combination term (term 1 and 2). It follows, therefore, that all terms of Equation 6-78 have one single common factor, which is $a'b'c'd'$. The resulting modified function therefore becomes:

$$Z = (a + b + a'b'c'd')(a + c + d + a'b'c'd')(b + c + d + a'b'c'd') \tag{6-79}$$

This function can be represented by the five-element implementation shown in Figure 6-36.

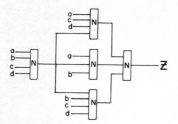

Figure 6-36 Minimal implementation of Example 6-6.

Again, for the application of NAND logic elements, the above method may be applied; however, a disjunctive minimal expression should be provided.

6-9 COMBINATIONAL NETWORK HAZARDS

The existence of a hazard in logic systems is often the cause of unpredictable operations. Although the elimination of such hazards in a logic system constitutes a serious obstacle in the implementation of these systems, it must be accomplished. In order to eliminate hazards, the designer must be aware of the conditions under which hazards occur.

Generally speaking, hazards are the result of imperfections in the actual conditions that are not accounted for in the description of a pure mathematical system. In fact, fluid components are far from perfect—they are relatively slow, and their speed of response is unpredictable. Unpredictability of the response of a logic element plays an important role in the creation of a hazard. In the mathematical model, a fluid logic component is assumed to have the output value of either 1 (ON) or 0 (OFF). At no time are intermediate conditions considered and the switching time is assumed to be zero.

In order to illustrate the difference between the ideal (mathematical) response and the actual, note the curves in Figure 6-37. The ideal responses of various functions are shown in the first column, while three different actual situations are presented in adjacent columns. This figure shows that the responses of the actual fluid logic devices are not perfect step functions as would be expected theoretically and that the switching speed might differ from one element to the other. As a consequence, glitches may develop in the outputs of certain downstream elements as illustrated in the functions for $(X_1 + X_2')$ and $X_1 X_2'$.

Hazards can be defined as a possible or actual deviation of the required response of a switching circuit or part of a switching circuit which occurs during the transition between two states (which may be input states or machine states). As the output of a combinational network is solely dependent upon the states of the inputs, the disagreement from the intended response is temporary, and this may or may not influence the main operation of the logic system.

Generally, hazards in combinational logic networks can be classified as hazards that are caused by transitions between two adjacent or nonadjacent terms (when portrayed in the Karnaugh map). Such hazards can be categorized into two major classes—static type and dynamic type.

COMBINATIONAL LOGIC DESIGN

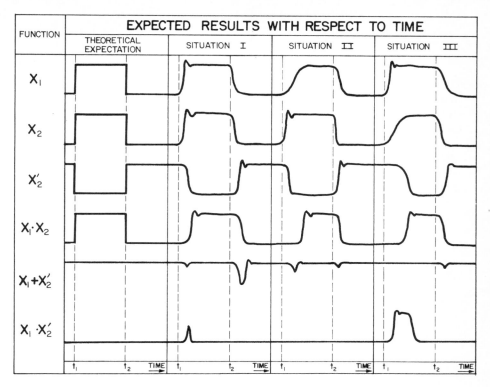

Figure 6-37 Various responses of fluid elements.

Static-type hazards are said to occur when the intended response is to remain unchanged. In this case, the "glitches" resulting from this hazard are characterized by an even number of changes within the output—see Figure 6-38(a). The dynamic hazard, on the other hand, is characterized by an odd number of changes in the output, as illustrated in Figure 6-38(b). The static hazard is best described by the Karnaugh map illustrated in Figure 6-39 together with its network diagram. The

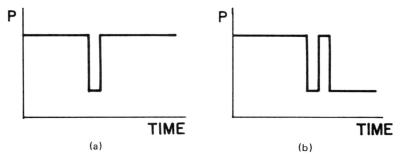

Figure 6-38 Forms of hazards: (a) Static-type hazard; (b) dynamic-type hazard.

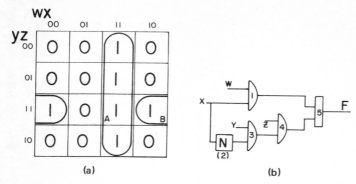

Figure 6-39 The static-1 hazard: (a) Karnaugh map; (b) circuit representation.

equation represented by the Karnaugh map is:

$$F = WX + X'YZ \tag{6-80}$$

In the network diagram shown in Figure 6-39, assume that the machine was initially in state A as indicated in the Karnaugh map at $WXYZ = 1111$. The change of variable X from 1 to 0 causes the machine to traverse to state B, which has the same output condition as state A (i.e., ON). Theoretically and ideally, the output F should remain ON during the transition; however, it is entirely possible that the delay encountered in the activation of the NOT valve (2) is longer than the delay in the deactivation of the AND valve (1), causing a temporary deactivation of the OR valve (5). As the above hazard involves a change of states in the 1-condition of the output, it is often referred to as the static-1-hazard. This hazard can be eliminated by including an implicant that is common to states A and B, which is the term WYZ. The reader may verify from the network in Figure 6-40 that the added term prohibits the OR valve from temporarily discharging the output F.

When a circuit is implemented conjunctively, a similar situation can occur, but during the OFF condition of the output. Consider the conjunctive expression as follows:

$$F = (W + X')(X + Z') \tag{6-81}$$

The disjunctive Karnaugh map of the expression and its circuit implementation can

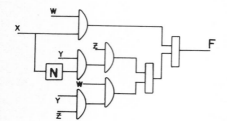

Figure 6-40 Static hazard-free network.

COMBINATIONAL LOGIC DESIGN

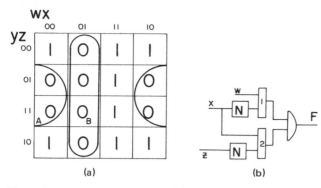

Figure 6-41 The static-0 hazard: (a) Karnaugh map; (b) circuit representation.

be observed in Figure 6-41. It can be verified that a temporary activation may occur if the OR valve (1) does not drain fast enough before the activation of the OR valve (2), during the transition caused by the change in X. As this irregularity involves a change in the OFF condition of the output, it is known as the static-0-hazard. Similarly, as in the static-1-hazard, the elimination of this hazard is accomplished by the inclusion of an additional implicate that is common to both terms, in this case $(W + Z')$. The hazard-free representation of F hence becomes:

$$F = (W + X')(X + Z')(W + Z') \qquad (6\text{-}82)$$

As a general rule, elimination of static hazards (0-type or 1-type), requires the inclusion of critical PI's (prime implicants or implicates) of the function. PI's are selected such that every pair of cubes in the Karnaugh map that represents 1's (if the expression is disjunctive) or 0's (if conjunctive) is "covered" by a prime implicant subcube or by a complement of the prime implicate subcube, as has been demonstrated by the above situation.

An almost similar situation occurs when multiple input changes are permitted to occur in a machine. For example, consider the Karnaugh map shown in Figure 6-42 with its circuit implementation. By noting every single variable value change at every location in the map, it can be seen that the network represented by this map is free from static hazards. However, if the inputs are permitted to change simultaneously from state $(ABCD = 1101)$ to state $(ABCD = 1011)$, another type of hazard prevails which is known as the static-1 logic hazard. It can be verified by circuit implementation that a temporary deactivation is really possible during such transitions. The procedure of elimination of these types of hazards is similar to that of the conventional static hazards. In relation to this, Eichelberger has shown that the inclusion of all PI's of a function is required for the elimination of both static-1 and static-0 logic hazards.

A function hazard is a completely different type of hazard. It is a hazard that occurs due to the function itself, and this occurs only when multiple-input-type simultaneous transitions are permitted in the system. In particular, these types of

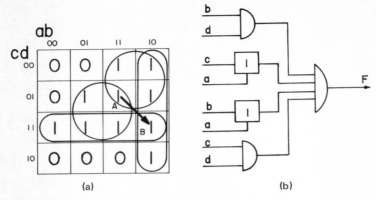

Figure 6-42 Static-1 logic hazard: (a) Karnaugh map; (b) circuit representation.

hazards prevail when no restrictions are made upon the simultaneous changes of the inputs and when more than one term exists in the expression. As an illustration, the Karnaugh map in Figure 6-43 is considered where a static-1 function hazard is shown. The intended change is shown in the map by the bold arrow, while the dashed arrows indicate the possible "paths" that may actually occur during the transition. It is therefore quite obvious that no mathematical cure can be made to correct or eliminate this type of hazard. It can be noted that for the static-1-type hazards (both logic and function), a similar static-0-hazard exists, mainly when the networks are constructed conjunctively. Furthermore, it is important to note that the prevalence of a static-type hazard always implies the existence of a relating dynamic-type hazard. For example, the "dynamic function hazard" in the Karnaugh map of Figure 6-44 exists due to the existence of a static-function hazard. It is worth noting also that the elimination of the static-type hazards will automatically eliminate dynamic hazards.

Although hazards can be eliminated mathematically, it is always possible for hazards to be reintroduced into the outputs by the hardware itself. For example, consider the following static hazard-free equation:

$$F = AB + BC \tag{6-83}$$

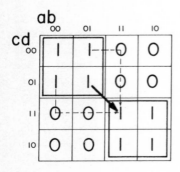

Figure 6-43 Static-1 function hazard.

COMBINATIONAL LOGIC DESIGN   171

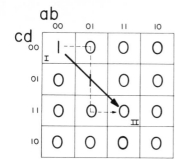

Figure 6-44 Dynamic function hazard.

Assume that no simultaneous changes in the inputs are permissible, and therefore, no function hazards can prevail. If, for instance, a three-way OR valve is used for the network representation, a hazard has been introduced into the output of the network. To further demonstrate this phenomenon, consider the three-way OR valve shown in Figure 6-45. The actual mathematical representation of this valve is:

$$Z = F_1 + F_1'F_2 \tag{6-84}$$

The simplification of the above equation is misleading, since the output Z is not $F_1 + F_2$ as Boolean algebra would indicate. Implementation of Equation 6-83 with this valve and using it as a regular or true OR will actually result in the implementation of the following equation:

$$\begin{aligned} F &= AB + (AB)'BC = \\ &\quad AB + (A' + B')BC = \\ &\quad AB + A'BC \end{aligned} \tag{6-85}$$

This representation is obviously hazardous, as there is a pair of Karnaugh cubes that is not covered by a subcube or an implicant, as can be observed in Figure 6-46.

The above example shows that it is not always advantageous to eliminate static or logic hazards by pure algebraic means. In these cases, the addition of prime implicants in the expression not only tends to increase the complexity of the circuit, but the effort may be fruitless, as shown in the above example. It is also a fact that no algebraic means has been developed that can eliminate function

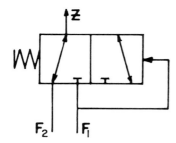

Figure 6-45 A three-way OR valve.

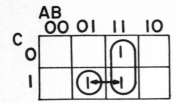

Figure 6-46 Karnaugh map for Equation 6-85.

hazards. To counteract this void in the technology, the use of various "smoothing" or damping concepts has proven effective for all types of hazards, both static or dynamic. Smoothing can be performed in two ways—by an R-C network and by using an inherent delay of a logic element. The R-C network can be observed in Figure 6-47(a), while its response may be seen in Figure 6-47(b). Note that the smoothened output is actually slower and its delay is dependent upon the adjustments of the R-C network. The concept of using the inherent delay of a logic element is achieved by using the AND valve, as shown in Figure 6-47(c). Here, the AND valve has become a logic valve that is often called a "yes" valve or a normally closed relay valve, the response of which can be observed in Figure 6-47(d).

6-10 COMPUTER-AIDED SYNTHESIS

Although the solution of example problems in the earlier sections of this chapter has been simple and straightforward, manual synthesis is almost impractical when large networks are encountered. On the other hand, computer programs capable of performing the desired function are generally lacking. For example, the development of a computer program which would consider all types of elements discussed in Chapter 4 is far from reality at this time. Even the existence of programs that can perform simple tasks for the designer is limited. Some of the computer programs which the authors feel are important in the area of combinational logic synthesis include the following:

1. The single-output minimization and static hazard elimination program.
2. The single-output complementation program.
3. The multiple-output network simplification program.
4. The multiple-output, limited fan-in/fan-out NAND network synthesis program.

The first program, i.e., the computer program for simplification of normal Boolean expressions (TAB II) is written in ANS Fortran IV. As discussed briefly in Chapter 2, the program is capable of simplifying equations having up to eighty variables. Although it is advisable to place a limit on the inputs to about thirty terms, the actual limitation of the program is the number of implicants/implicates that are generated during the execution of the program, which is limited to seventy-five terms. A total storage array of 19005 array locations are used by the program, and as ANS Fortran requires four bytes (32 bits) to store each array element, a 76-kByte storage is required (total of 608 kilobits). As the program leaves a large number of bits unused (only two bits are actually required to store

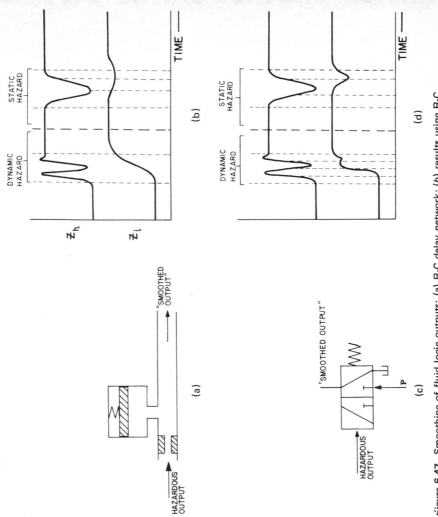

Figure 6-47 Smoothing of fluid logic outputs: (a) R-C delay network; (b) results using R-C delay; (c) AND valve smoothing; (d) results using AND valve smoothing.

each array element), this program can be further expanded to tolerate even larger problems with less storage requirements.

TAB II is not only capable of simplifying equations containing "don't care" entries, but is even capable of generating the needed but unspecified PI's of the "don't cares." Basically, TAB II generates only the necessary PI's of the ON and the "don't care" entries by complementation of the OFF entries. Another feature offered by TAB II is its ability to produce the simplest, static-hazard-free network equations upon request. TAB II accepts a Boolean expression in numerical format, which is entered in array form. In this array, 0's, 1's, and 2's are used to represent the complemented, the uncomplemented, and the indeterminate conditions of each variable, respectively. For example, for the expression:

$$Z = AB'D'E'F'H'IJ + A'B'CD'EHI + AB'DE'F'G'H'IJ + \\ A'BCDE'HI + ABDF'GH'IJ + A'BCDGHI + ABDF'G'H'IJ + \\ A'BCDG'HI \qquad (6\text{-}86)$$

The input array to the program is:

$$Z = \begin{array}{c} \begin{array}{cccccccccc} A & B & C & D & E & F & G & H & I & J \end{array} \\ \begin{bmatrix} 1 & 0 & 2 & 0 & 0 & 0 & 2 & 0 & 1 & 1 \\ 0 & 0 & 1 & 0 & 1 & 2 & 2 & 1 & 1 & 2 \\ 1 & 0 & 2 & 1 & 0 & 0 & 0 & 0 & 1 & 1 \\ 0 & 1 & 1 & 1 & 0 & 2 & 2 & 1 & 1 & 2 \\ 1 & 1 & 2 & 1 & 2 & 0 & 1 & 0 & 1 & 1 \\ 0 & 1 & 1 & 1 & 2 & 2 & 1 & 1 & 1 & 2 \\ 1 & 1 & 2 & 1 & 2 & 0 & 0 & 0 & 1 & 1 \\ 0 & 1 & 1 & 1 & 2 & 2 & 0 & 1 & 1 & 2 \end{bmatrix} \end{array} \qquad (6\text{-}87)$$

If both the absolute minimum and the minimal static-hazard-free equations are requested, then the solutions are returned by means of two arrays, which would resemble the following:

$$Z_{\min} = \begin{array}{c} \begin{array}{cccccccccc} A & B & C & D & E & F & G & H & I & J \end{array} \\ \begin{bmatrix} 0 & 1 & 1 & 1 & 2 & 2 & 2 & 1 & 1 & 2 \\ 1 & 1 & 2 & 1 & 2 & 0 & 2 & 0 & 1 & 1 \\ 0 & 0 & 1 & 0 & 1 & 2 & 2 & 1 & 1 & 2 \\ 1 & 0 & 2 & 0 & 0 & 0 & 2 & 0 & 1 & 1 \\ 1 & 0 & 2 & 2 & 0 & 0 & 0 & 0 & 1 & 1 \end{bmatrix} \end{array} \qquad (6\text{-}88)$$

$$Z_{\text{h.f.}} = \begin{array}{c} \begin{array}{cccccccccc} A & B & C & D & E & F & G & H & I & J \end{array} \\ \begin{bmatrix} 0 & 1 & 1 & 1 & 2 & 2 & 2 & 1 & 1 & 2 \\ 1 & 1 & 2 & 1 & 2 & 0 & 2 & 0 & 1 & 1 \\ 0 & 0 & 1 & 0 & 1 & 2 & 2 & 1 & 1 & 2 \\ 1 & 0 & 2 & 0 & 0 & 0 & 2 & 0 & 1 & 1 \\ 1 & 0 & 2 & 2 & 0 & 0 & 0 & 0 & 1 & 1 \\ 1 & 2 & 2 & 1 & 0 & 0 & 0 & 0 & 1 & 1 \end{bmatrix}^{*} \end{array} \qquad (6\text{-}89)$$

*Hazard-eliminating term.

COMBINATIONAL LOGIC DESIGN

When "don't care" terms are available, they are presented in a form as illustrated by the following array:

$$Z_{d.c.} = \begin{bmatrix} A & B & C & D & E & F & G & H & I & J \\ 1 & 0 & 2 & 1 & 2 & 0 & 1 & 0 & 1 & 1 \\ 0 & 0 & 1 & 2 & 0 & 2 & 0 & 1 & 1 & 2 \\ 1 & 1 & 2 & 0 & 1 & 0 & 2 & 0 & 1 & 1 \\ 0 & 1 & 1 & 0 & 1 & 2 & 2 & 1 & 1 & 2 \\ 0 & 0 & 1 & 1 & 1 & 2 & 1 & 1 & 1 & 2 \end{bmatrix} \quad (6\text{-}90)$$

then, if the hazard-free equation is requested, the following array is returned by subroutine TAB II:

$$Z_{h.f.} = \begin{bmatrix} A & B & C & D & E & F & G & H & I & J \\ 0 & 1 & 1 & 1 & 2 & 2 & 2 & 1 & 1 & 2 \\ 0 & 2 & 1 & 0 & 1 & 2 & 2 & 1 & 1 & 2 \\ 1 & 1 & 2 & 1 & 2 & 0 & 2 & 0 & 1 & 1 \\ 1 & 0 & 2 & 2 & 0 & 0 & 2 & 0 & 1 & 1 \\ 1 & 2 & 2 & 1 & 0 & 0 & 2 & 0 & 1 & 1 \end{bmatrix} * \quad (6\text{-}91)$$

*Hazard-eliminating term.

Note that this result may contain static hazards; however, these hazards would occur only in cube-transitions that involve "don't care" cubes. (TAB II assumes that hazards under these circumstances are tolerable.)

In other problems, however, situations may exist where the OFF terms are available to the designer. In such situations, the designer is free to assume that the unspecified entries are all "don't cares." For example, if the following OFF array is supplied:

$$Z' = \begin{bmatrix} A & B & C & D & E & F & G & H & I & J \\ 1 & 0 & 2 & 0 & 1 & 0 & 2 & 0 & 1 & 1 \\ 0 & 1 & 1 & 0 & 0 & 2 & 2 & 1 & 1 & 2 \\ 1 & 0 & 2 & 2 & 1 & 0 & 0 & 0 & 1 & 1 \\ 0 & 0 & 1 & 2 & 0 & 2 & 1 & 1 & 1 & 2 \\ 1 & 1 & 2 & 0 & 0 & 0 & 2 & 0 & 1 & 1 \\ 0 & 0 & 1 & 1 & 1 & 2 & 0 & 1 & 1 & 2 \end{bmatrix} \quad (6\text{-}92)$$

then TAB II will generate the required PI's of the "1" and the "-" entries such that the minimized result is obtained:

$$Z_{min} = \begin{bmatrix} A & B & C & D & E & F & G & H & I & J \\ 2 & 1 & 2 & 1 & 2 & 2 & 2 & 2 & 2 & 2 \\ 1 & 0 & 2 & 2 & 0 & 2 & 2 & 2 & 2 & 2 \\ 0 & 2 & 2 & 0 & 1 & 2 & 2 & 2 & 2 & 2 \end{bmatrix} \quad (6\text{-}93)$$

The complementation program (subroutine NOR) is also written in the ANS Fortran IV language. It is designed to complement equations that occupy at most 1000-array locations; and an output array of 5000 locations is provided for the program. It should be noted that the actual limitation on this program is the output array—the diverging characteristics of the complementation operation often require an enormous amount of storage to compile the PI's of the complement. This program requires only 7202 array locations, which in ANS Fortran IV is magnified to an approximately 29-kBytes array storage (232 kilobits).

Subroutine NOR finds various uses in the area of logic, such as the generation of conjunctive equations from disjunctive equations, and vice versa; also, the generation of the still unspecified states of a system. For example, if the following disjunctive equation is supplied (in a numerical array as in subroutine TAB II):

$$Z = A\,B'\,C + A'\,C' = \begin{matrix} A & B & C \\ \end{matrix} \begin{bmatrix} 1 & 0 & 1 \\ 0 & 2 & 0 \end{bmatrix} \tag{6-94}$$

and if the PI's of the conjunctive equation are requested, then NOR will return the following array to the user:

$$Z = \begin{matrix} A & B & C \\ \end{matrix} \begin{bmatrix} 0 & 0 & 2 \\ 0 & 2 & 1 \\ 1 & 2 & 0 \\ 2 & 0 & 0 \end{bmatrix} \tag{6-95}$$

which is interpreted as:

$$Z = (A' + B')(A' + C)(A + C')(B' + C') \tag{6-96}$$

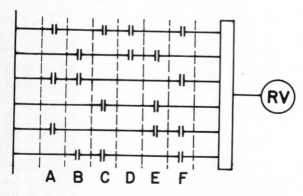

Figure 6-48 Relay diagram.

COMBINATIONAL LOGIC DESIGN 177

Closely related to this operation, NOR is capable of obtaining all possible cut sets of a relay network if the associated tie-sets are given. (A cut-set is a set of relays that insures, when closed, that the output is zero. A tie-set, on the other hand, is a set of relays that, when open, insures the activation of the output regardless of the conditions of the other relays.) A typical relay diagram is shown in Figure 6-48. In this diagram, there are six relays that are actuated by the input signals $A, B, C, D \ldots F$. Each of the horizontal branches represents a tie-set, while the cut-sets of the network are to be generated. The input array for the subroutine NOR can be constructed by inserting 1's wherever a relay is located, and 2's if otherwise. Hence, the array is as follows:

$$\begin{array}{cccccc} A & B & C & D & E & F \end{array}$$
$$\begin{bmatrix} 1 & 2 & 1 & 1 & 2 & 1 \\ 2 & 1 & 2 & 1 & 1 & 2 \\ 1 & 1 & 2 & 2 & 2 & 1 \\ 2 & 2 & 1 & 2 & 1 & 2 \\ 1 & 2 & 2 & 2 & 1 & 1 \\ 2 & 1 & 1 & 2 & 2 & 1 \end{bmatrix} \quad (6\text{-}97)$$

Executing the program without complementation (as if requesting a conjunctive form from a disjunctive form), the following array which reveals all of the shortest cut-sets of the diagram:

$$\begin{array}{cccccc} A & B & C & D & E & F \end{array}$$
$$\begin{bmatrix} 2 & 2 & 2 & 2 & 1 & 1 \\ 2 & 1 & 1 & 2 & 2 & 1 \\ 2 & 2 & 1 & 1 & 2 & 1 \\ 1 & 1 & 1 & 2 & 2 & 2 \\ 1 & 1 & 2 & 2 & 1 & 2 \\ 1 & 2 & 1 & 1 & 2 & 2 \\ 1 & 2 & 1 & 2 & 1 & 2 \\ 2 & 1 & 1 & 2 & 1 & 2 \\ 2 & 1 & 2 & 1 & 1 & 2 \end{bmatrix} \quad (6\text{-}98)$$

In relation to the second use of the NOR program (generating the still unspecified states of the system), the previously assessed program, TAB II, utilizes the same algorithm to generate the desired PI's by complementation of the OFF terms. As the complementation operation is divergent, in TAB II not all of the PI's are generated during the execution. Only "strong" candidates are generated, which, for example, can be demonstrated by the complementation of Equation 6-92. The prime implicant of the "1" and "-" entries are shown in Table 6-3, while the subroutine TAB II only generates the entries as indicated by the *-marked entries in Table 6-3. At this point only combinational logic synthesis programs for single-output networks have been discussed, but networks that have multiple outputs can

Table 6-3 The Complement of the OFF Terms of Equation 6-92

A	B	C	D	E	F	G	H	I	J		A	B	C	D	E	F	G	H	I	J	
2	2	2	2	2	2	2	2	0	2		2	2	0	1	2.2	1	2	2	2	
2	1	2	1	2	2	2	2	2	2	*	2	0	0	2	0	2	2	2	2	2	
2	1	2	2	1	2	2	2	2	2		2	2	2	1	0	2	2	0	2	2	*
0	2	0	2	2	2	2	2	2	2		2	2	2	1	1	2	1	2	2	2	
0	2	2	2	2	2	2	0	2	2		2	2	2	1	2	2	1	0	2	2	
1	2	2	2	2	1	2	2	2	2		2	2	2	1	2	1	2	0	2	2	
1	2	2	2	2	2	2	1	2	2		2	2	2	1	2	2	2	0	2	0	
1	2	2	2	2	2	2	2	2	0		2	0	2	2	0	2	2	0	2	2	*
2	2	0	2	2	1	2	2	2	2		2	2	2	2	0	1	2	0	2	2	
2	2	0	2	2	2	2	1	2	2		2	2	2	2	0	2	2	0	2	0	
2	2	0	2	2	2	2	2	2	0		2	2	2	0	1	1	2	2	2	2	
2	2	2	2	2	1	2	0	2	2		2	2	2	2	1	1	1	2	2	2	
2	2	2	2	2	2	2	0	2	0		2	2	2	2	1	1	2	0	2	2	
0	2	2	0	1	2	2	2	2	2	*	2	2	2	2	1	2	1	1	2	2	
0	2	2	2	1	2	1	2	2	2		2	2	2	2	1	2	1	2	2	0	
0	0	2	0	2	2	0	2	2	2		2	2	2	0	1	2	2	1	2	2	*
2	0	2	2	0	2	0	2	2	2	*	2	2	2	0	1	2	2	2	2	0	
2	2	2	1	0	2	0	2	2	2	*	2	2	2	1	2	2	0	2	2	0	
1	2	2	1	0	2	2	2	2	2	*	2	0	2	0	2	1	0	2	2	2	
1	2	2	1	2	2	1	2	2	2		2	0	2	0	2	2	0	1	2	2	
1	0	2	2	0	2	2	2	2	2	*	2	0	2	0	2	2	0	2	2	0	
2	2	0	1	0	2	2	2	2	2												

*Term generated by TAB II.

be synthesized by these same programs, though each output has to be synthesized individually. In this regard, it is known that synthesizing a network as a whole often produces greatly simplified results, especially when a large number of outputs exists. To implement this, the multiple output simplification (MOMIN) has been developed, written in the Fortran V language for the Univac 1108 digital computer. The program utilizes approximately 19200 array locations and limits the array size that is used for the storing of the implicants to about 2400 locations including the outputs (i.e., Input + Output variables × States must be less than 2400). However, as the program stores 12-array elements in one 36-bit word, it is efficient in storage requirements—only 1620 words (51.84 kilobits) are needed. As with the other computer programs, MOMIN accepts a problem description in an array format. The truth table of the problem statement is described in the array, using x's for indeterminate variables and d's for "don't care" outputs. Hence, the following array:

$$
\begin{array}{ccccc}
a & b & c & Z_1 & Z_2 \\
\end{array}
$$

$$
\begin{bmatrix}
0 & 0 & 0 & 1 & 1 \\
0 & 0 & 1 & d & 1 \\
0 & 1 & 1 & 1 & 0 \\
0 & 1 & 0 & 1 & 1 \\
1 & 1 & 0 & 1 & d \\
1 & 1 & 1 & 0 & 0 \\
1 & 0 & 1 & 0 & d \\
1 & 0 & 0 & 1 & 1
\end{bmatrix}
\quad (6\text{-}99)
$$

COMBINATIONAL LOGIC DESIGN

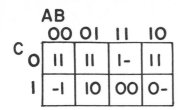

Figure 6-49 Karnaugh map of Equation 6-99.

represents the output functions described by the Karnaugh map of Figure 6-49. The program returns a matrix in the same format:

$$a \quad b \quad c \quad Z_1 \quad Z_2$$
$$\begin{bmatrix} x & x & 0 & 1 & 1 \\ 0 & x & x & 1 & 0 \\ x & 0 & x & 0 & 1 \end{bmatrix} \quad (6\text{-}100)$$

In this output matrix, however, it should be noted that the interpretation is slightly different: $ABCZ_1Z_2 = 0xx10$ does not mean that Z_2 is totally independent from $ABC = 0xx$. This representation only indicates that the output Z_1 is "connected" to a circuit that produces $ABC = 0xx$, while Z_2 is not.

Another powerful program is the NAND synthesis program for multiple-output networks. It is programmed in Fortran V on the CDC-6400 computer and limits the problem to about fifty states and seventy-five variables (input + output variables). In total, the program utilizes an array of about 284,320 locations, and by using three bits of computer core to store information pertaining to each location, only 14,216 60-bit words (or 853 kilobits) are required to store the entire working array of the program. The input to the program is essentially the same as MOMIN—the truth table of the control system must be supplied. The difference lies in the output—the program supplies the complete description of the connections necessary for the implementation upon NAND networks. Other additional features offered by this program are the fan-in and the fan-out limits that can be posed upon the problem, and also the weighing criteria for the solution of fan-in and fan-out problems. This weighing criteria determines the relative importance of the fan-in and fan-out problems. For example, a fan-in weight twice as much as a fan-out weight indicates that it is twice as important to solve fan-in problems than fan-out problems.

As an illustration of the ability of this program, consider the truth table shown in Table 6-4. By limiting the fan-in and fan-out of the available elements to two and setting the weighting functions of the fan-in and fan-out to be equal, the computer output shown in Table 6-5 is obtained. In this computer output, the interpretation should be made as follows:

1. 8 NAND A2, 5 should be interpreted as NAND No. 8 has its inputs connected to input A2 and NAND valve No. 5.
2. Z1 21 NAND 12, 19 should be interpreted as Z1 results from NAND No. 21.

Table 6-4 Truth Table

A_1	A_2	A_3	A_4	Z_1	Z_2	Z_3	Z_4
0	x	0	0	1	1	0	0
0	0	0	1	0	1	0	0
0	0	1	1	0	0	0	0
0	0	1	0	0	d	0	0
0	1	0	1	0	1	0	1
0	1	1	1	0	0	1	0
0	1	1	0	0	0	d	0
1	1	0	0	1	d	0	0
1	x	0	1	0	0	0	0
1	1	1	x	1	1	1	1
1	0	x	0	1	1	0	0
1	0	1	1	0	1	1	1

PROBLEMS

1 By using the the conventional approach, the following equations are to be implemented using a "mixed logic" form. Assume that only two-input AND's, OR's, and NOT's are available.

  a $W = A'BCD + ABD' + AC'D' + AB'C'$

Table 6-5 Computer Results

	1	NAND	A_1, A_3
	2	NAND	$-A_1, -A_3$
	3	NAND	A_2, A_4
	4	NAND	A_3
	5	NAND	1
	6	NAND	2
	7	NAND	3
	8	NAND	$A_2, 5$
	9	NAND	$A_4, 5$
	10	NAND	6, 7
	11	NAND	$A_1, -A_4$
	12	NAND	$-A_1, 4$
	13	NAND	$A_3, 7$
	14	NAND	$-A_1, 4$
	15	NAND	8, 9
	16	NAND	8, 11
	17	NAND	9, 11
	18	NAND	15
	19	NAND	16
	20	NAND	17
Z_1	21	NAND	12, 19
Z_2	22	NAND	14, 20
Z_3	23	NAND	13, 18
Z_4	24	NAND	10, 18

COMBINATIONAL LOGIC DESIGN

  b $X = BCD' + ACD + A'BC'D + A'CD'$
  c $Y = AB'C'D' + A'B'C + A'CD$
  d $Z = BC'D' + ACD' + A'D$
  If INHIBITOR elements were available, how does this affect the network construction?

2 Network functions represented by the truth tables of Table P6-2 are to be implemented using two-input AND, OR, and INHIBITOR elements. First use the conventional method without incorporating the "don't cares"; then compare the network constructions with the results when the "don't cares" are effectively utilized. Assume that only uncomplemented input signals are available.

Table P6-2a

abcd	Z_1	Z_2
0000	0	0
0100	0	1
0001	1	1
0010	1	0
1000	–	0
1111	1	–

Table P6-2b

abcd	Z_1	Z_2	Z_3
0000	0	0	1
010–	1	0	0
110–	0	1	–
–110	1	1	–
0011	–	0	0
1111	–	1	–
1011	0	–	1

3 Implement Problem 1 and 2 using:
  a Unlimited fan-in AND and OR elements.
  b Using all types of elements available as discussed in previous chapters, by selecting the "correct" type of element for each particular assignment, try to obtain a most simplified circuit.

4 By taking the complements of the problem descriptions of Problems 1 and 2, obtain the simplest representation using:
  a Two-input elements.
  b Multi-fan-in elements.

5 The truth tables of Table P6-5 represent networks that are to be implemented by means of all available fluid elements. Assuming that only uncomplemented input variables are available, use complementation techniques to obtain the "minimal" circuit representation.

Table P6-5a

ABCD	Z_1	Z_2	Z_3
0000	1	1	0
0100	0	1	1
0010	1	0	1
0001	1	–	1
1000	0	0	–
1001	1	1	1
0110	0	0	1

Table P6-5b

ABCD	X	Y	Z
000–	0	1	0
0100	1	1	0
–010	0	0	1
1110	0	1	1
1–00	0	1	0
01–1	1	0	1
1–01	0	0	0

Table P6-5c

ABCDE	W	X	Y	Z
00000	0	0	0	1
01000	0	1	0	1
10000	1	0	–	1
00100	0	1	0	1
00010	0	0	1	0
11010	1	–	0	0
00001	0	0	0	0
10001	1	1	–	1

Table P6-5d

ABCD	Z_1	Z_2	Z_3	Z_4	Z_5
0–00	0	1	0	1	0
1100	–	1	–	1	0
11–1	1	1	0	1	0
00–1	1	–	0	–	1
1000	0	0	1	0	1
01–1	1	0	–	1	0
0010	0	0	0	0	1

6 Using factoring techniques, simplify, if possible, the circuit of Problems 2 and 5.

7 The following network equations are to be implemented by the following means:
   a Two-input AND, OR, and INHIBITOR elements.
   b Multi-input AND/OR elements (and a one-input NOT).
   c All types of logic elements as discussed in Chapter 4.
   The equations are:
   $Z_1 = a'bc'd + abc' + abdf$
   $Z_2 = abc'd + bc'f + ab'f$
   $Z_3 = bc'f + b'df' + acd$
   $Z_4 = a'bc' + bc'df + af'$
   Try to minimize the number of elements involved in the network by utilizing any of the available approaches.

8 Suppose that the network equation of Problem 7 is to control a hydraulic system. In order to avoid interfacing problems, it is suggested that hydraulic logic elements be used. Only spool-type three-way and four-way valves are to be utilized to satisfy the logic functions.
   a Derive the output equations of all types of three-way and four-way valves, such as spring-return or spring-centered valves; open-center or blocked-center valves, etc.
   b From the equations identify the possible applications of each valve, individually or in series.
   c Using all combinational synthesis techniques discussed in this chapter, obtain the minimal circuit.

9 Using multiple-terminal synthesis, obtain the minimal networks for the networks of Problem 5. Utilize unlimited fan-in logic elements.

10 Implement the networks depicted by Problems 2 and 5 by using four-input NOR logic elements only.

11 Solve the same problems indicated in Problem 10 using NAND logic elements having a fan-in number of four.

12 Implement the networks represented by the truth tables of Table P6-5 by using:
   a Unlimited fan-in/fan-out NOR logic elements.
   b Unlimited fan-in/fan-out NAND logic elements.
   c INHIBITOR elements only.

Minimize the number of stages used in order to optimize the speed of the network.

13 Implement the following conjunctive equations (shown in numerical array form) using three-level NOR logic synthesis:

a $\begin{array}{cccc} a & b & c & d \end{array}$
$$\begin{bmatrix} 1 & 1 & 1 & 0 \\ 1 & 0 & 1 & 1 \\ 0 & 1 & 1 & 1 \\ 1 & 1 & 0 & 1 \end{bmatrix}$$

b $\begin{array}{ccccc} a & b & c & d & e \end{array}$
$$\begin{bmatrix} 1 & 1 & - & 0 & - \\ 1 & 1 & 0 & - & - \\ 1 & 1 & - & - & 0 \\ 1 & 0 & 1 & 1 & 1 \\ 0 & 1 & 1 & 1 & 1 \end{bmatrix}$$

c $\begin{array}{cccc} a & b & c & d \end{array}$
$$\begin{bmatrix} 1 & - & - & 0 \\ 1 & - & 0 & - \\ 1 & 0 & - & - \\ 0 & 1 & 0 & 1 \\ - & 1 & 1 & 0 \\ 1 & 1 & - & 1 \end{bmatrix}$$

14 Implement the following network equations using three-level NOR logic synthesis:
 a $Z = ae + b'e + ce + d'e' + ace' + ab + a'bc$
 b $Z = ab + ac + ad + ae + be + ce + de + bd + cd$

15 From the following equations, develop the minimal static (1 or 0) hazard-free equations:
 a $W = a'b'c'd' + ab'd' + b'cd' + ad + a'd' + a'bd$
 b $X = abd' + bcd + a'c'd + a'b'd + dc'$
 c $Y = (a + b' + d)(a + c' + d')(a' + b)(a + b + d)(b' + c + d)$
 d $Z = (a + b + d)(a + c + d')(a + b' + c' + d)(a' + b + c + d')(a' + b')$

16 For the equations of Problem 13, develop the logic hazard-free equations.

DEFINITIONS

Hazard An actual or potential network malfunction as a result of network delays or imperfections.

Implementation The act or process of representing an algebraic description of a network by means of hardware application.

Level or stage location of an element The number of elements through which the input signals of the element under consideration must transgress in order to reach the output.

Levels or stages in a network The largest or maximal stage location in a network.

Logically complete element or operator An element or operator that can satisfy all logic functions either by itself or with the conjunction of equivalent elements or operators.

Multiple-output prime implicant or implicate An implicant or implicate that represents one or more outputs and that is irreducible for the representation of the particular combination of outputs.

Synthesis The act or process of resolving a logic description and determining the best algebraic description.

REFERENCES

Abhyankar, S., "Minimal 'Sum of Products of Sums' Expressions of Boolean Expressions," *IRE Transactions on Electronic Computers*, Vol. EC-7, pp. 268–276, Dec. 1958.

———, "Absolute Minimal Expressions of Boolean Functions," *IRE Professional Group on Electronic Computers, Transactions*, Vol. 7-8, pp. 3-8, March 1959.

Bartee, T. C., "Computer Design of Multiple-Output Networks," *IRE Transactions on Electronic Computers*, Vol. C-10, No. 1, pp. 21-30, March 1961.

Chakrabarti, K. K., Choudhury, A. K., and Basu, M. S., "Complementary Function Approach to the Synthesis of Three-Level NAND Network," *IEEE Transactions on Electronic Computers*, Vol. C-19, No. 6, pp. 509-514, June 1970.

DeMoss, D. M., "Criteria for the Design of Fast, Safe, Asynchronous Sequential Fluidic Circuits," Ph.D. Dissertation, Oklahoma State University, Stillwater, Oklahoma, 1967.

Dietmeyer, D. L., *Logic Design of Digital Systems*. Boston: Allyn and Bacon, Inc., 1970.

———, and Schneider, P. R., "A Computer-Oriented Factoring Algorithm for NOR-Logic Design," *IEEE Transactions on Electronic Computers*, Vol. EC-11, No. 6, pp. 868-874, Dec. 1965.

———, and Su, Y. H., "Logic Design Automation of Fan-In Limited NAND Networks," *IEEE Transactions on Electronic Computers*, Vol. C-18, No. 1, pp. 11-22, Jan. 1969.

Eichelberger, E. B., "Hazard Detection in Combinational and Sequential Switching Circuits," Proceedings, Fifth Annual Symposium on Switching Theory and Logical Design, pp. 111-120, Nov. 1964.

———, "The Synthesis of Combinational Circuits Containing Hazards," Department of Electrical Engineering, Digital Systems Laboratory, Princeton University, Report No. 17, March 1962.

Ellis, D. T., "Synthesis of Combinational Logic with NAND and NOR Elements," *IEEE Transactions on Electronic Computers*, Vol. EC-14, No. 5, pp. 701-705, Oct. 1965.

Fitch, E. C., "Fluid Logic," Teaching Manual Published by The School Of Mechanical and Aerospace Engineering, Oklahoma State University, Stillwater, Oklahoma, 1966.

Gimpel, J. F., "The Minimization of TANT Networks," *IEEE Transactions on Electronic Computers*, Vol. EC-16, No. 1, pp. 18-38, Feb. 1967.

Givone, D. D., *Introduction To Switching Circuit Theory*. New York: McGraw-Hill Book Co., 1970.

Grasselli, A., and Gimpel, J. F., "The Synthesis of Three Level Logic Networks," Department of Electrical Engineering, Digital Systems Laboratory, Princeton University, Technical Report No. 49, May 1966.

Huffman, D. A., "The Design and Use of Hazard-Free Switching Networks," *Journal of Association of Computing Machinery*, Vol. 4, No. 1, pp. 47-62, Jan. 1957.

McCluskey, E. J., "Logical Design Theory of NOR Gate Networks With No Complemented Inputs," Proceedings, Fourth Annual Symposium on Switching Theory and Logical Design, pp. 137-148, 1963.

———, and Bartee, T. C., "A Survey of Switching Circuit Theory," Chapter V, in E. J. McCluskey (ed.), *Minimization Theory*. New York: McGraw-Hill Book Co., pp. 67-88, 1962.

———, and Schorr, H., "Essential Multiple Output Prime Implicants," Department of Electrical Engineering, Digital Systems Laboratory, Princeton University, Technical Report No. 23, April 1962.

Schneider, P. R., and Dietmeyer, D. L., "An Algorithm for Synthesis of Multiple Output Combinational Logic," *IEEE Transactions on Electronic Computers*, Vol. C-18, No. 2, pp. 117–128, Feb. 1969.

Slagle, J. R., Chang, C. L., and Lee, R. C. T., "A New Algorithm For Generating Prime Implicants," *IEEE Transactions on Electronic Computers*, Vol. C-19, No. 4, pp. 304–310, April 1970.

Su, Y. H., "Multiple Output NAND Networks," *Electronic Design*, No. 23, pp. 98–105, November 8, 1973.

——, and Dietmeyer, D. L., "Computer Reduction of Two-Level, Multiple Output Switching Circuits," *IEEE Transactions on Electronic Computers*, Vol. C-18, No. 1, pp. 58–63, Jan. 1969.

——, and Nam, C. W., "Computer Aided Synthesis of Multiple-Output Multilevel NAND Networks with Fan-In and Fan-Out Constraints," *IEEE Transactions on Electronic Computers*, Vol. C-20, No. 12, pp. 1445–1455, Dec. 1971.

Surjaatmadja, J. B., "A Computer Oriented Method for Boolean Simplification and Potential Hazard Elimination," Fluid Power Research Conference, Report No. R73-FL-2, Fluid Power Research Center, Oklahoma State University, Stillwater, Oklahoma, October 1973.

——, "A Generalized Method for Synthesizing Optimal Fluid Logic Networks," Fluid Power Research Conference, Report No. R73-FL-3, Fluid Power Research Center, Oklahoma State University, Stillwater, Oklahoma, October 1973.

——, "TAB II-Revised Program and Users Guide for the Simplification and Static Hazard Elimination of Colossal Boolean Expressions," Fluid Power Research Conference, Report No. R75-2, Fluid Power Research Center, Oklahoma State University, Stillwater, Oklahoma, October 1975.

——, "A Computer-Oriented Method For Complementing Boolean Expressions," Fluid Power Research Conference, Paper No. P75-59, Fluid Power Research Center, Oklahoma State University, Stillwater, Oklahoma, October 1975.

——, "Hazards in Fluid Logic Systems," Fluid Power Research Conference, Paper No. P75-57, Fluid Power Research Center, Oklahoma State University, Stillwater, Oklahoma, October 1975.

Torng, H. C., *Switching Circuits: Theory and Logic Design*. Boston: Addison-Wesley Publishing Company, Inc., 1972.

Zissos, D., *Logic Design Algorithms*. London: Oxford University Press, 1972.

——, and Duncan, F. G., "Boolean Minimization," *The Computer Journal*, Vol. 16, No. 2, pp. 174–179, March 1973.

Chapter 7

Sequential Logic Design–Classical Synthesis

7-1 LOGIC DESCRIPTION

Synthesis is the coupling of complex parts or elements to form a coherent system capable of satisfying a predesignated objective. In relation to fluid logic, the process consists of formulating a network of logic elements which functions according to a prescribed set of requirements. Hence, the fundamental goal in synthesizing any control network is to establish a correspondence between network inputs and outputs commensurate with the network specifications. Such specifications serve as the foundation for initiating a synthesis process and become uniquely reflected in the results of the synthesis.

The Primitive Flow Table (PFT) advanced by Huffman in 1954 for describing the fundamental mode of sequential machines continues to serve as the basis for the classical synthesis method of fluid logic. The fundamental mode refers to the operation of a machine (or control network of a machine) where the inputs never change unless the network itself is in a "stable" condition. Here, the term "stable" means that the condition or state of the inputs and memories is such that no change in the memory elements is required. The construction of the PFT was presented in Chapter 5.

SEQUENTIAL LOGIC DESIGN—CLASSICAL SYNTHESIS

The classical synthesis method for designing sequential logic systems has from its inception utilized the PFT as well as the synthesis methodology introduced by Huffman. This method has been continually improved over the years by investigators incorporating the most advanced design techniques which were compatible with the characteristic premise of the Huffman model. With such updating, the method continues to serve the important function of an uncompromising "referee" method for assessing the capability of all other methods. Although important, this appraisement function should never be construed as being subordinate to its great utility value in synthesizing fluid logic networks.

As indicated in Chapter 5, the form of the PFT as originally conceived by Huffman doubles in physical (column) size for each input added to the system. This proliferation of columns for large-scale networks is recognized as a serious limitation of the PFT and could easily jeopardize the future of the classical method if ignored. Fortunately, the advancement of the Logic Specification Chart (LSC) as presented in Chapter 5 and illustrated in Figure 7-1 should rebuff such objections and help perpetuate the classical method. This chart minimizes the number of columns, preserves the logic description capability of the PFT, and offers an ideal input format for computer programs.

The LSC is basically the same as the PFT in general features. The major difference is that only those inputs which actually occur or are considered are listed and then only in a random order. Instead of the outputs being recorded on the right side of the chart (restricting the promulgation of the input states), they are listed on the left side. The LSC will serve as the logic description for the classical synthesis method presented in this book. The reader should refer to Chapter 5 for further details on the construction of the LSC.

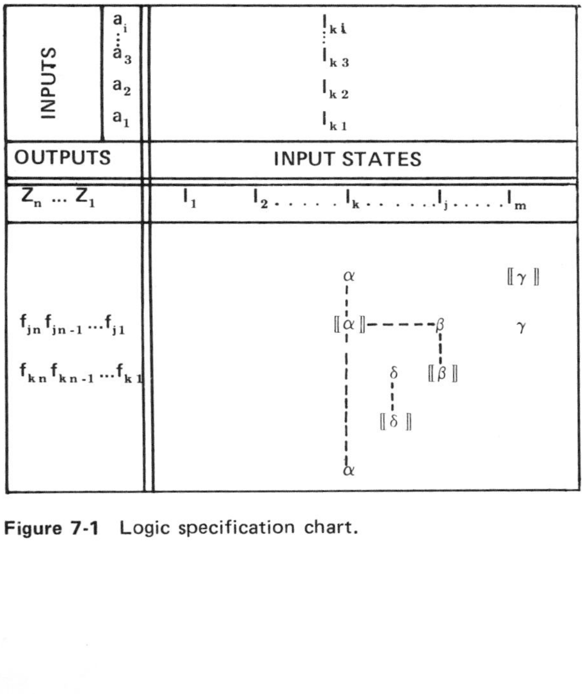

Figure 7-1 Logic specification chart.

7-2 STATE EQUIVALENCY

The Logic Specification Chart (LSC) represents a concise logic statement or description of the desired behavior of a sequential circuit in an unrefined form. In the process of its development, no effort is made to eliminate redundant machine states, as they are not always apparent. It should be realized, however, that redundant states normally lead to a more complex network and should therefore be minimized. One method for eliminating redundant states is discussed in this section and involves the recognition of equivalent states associated with the LSC.

Two stable states in the same column of the LSC are classified as equivalent if the following conditions exist:

1. The associated stable states are in the same column of the LSC.
2. The outputs are either identical for all input states, or where disagreement occurs, "don't care" (-) entries are involved.
3. The states in each column of both rows are either equivalent or disagree due to an unspecified (blank, or "-") entry.

To illustrate the establishment of equivalent states, consider the LSC shown in Figure 7-2. In this chart, states 1 and 7 are equivalent only if states 3 and 5 are equivalent (condition c). States 3 and 5 are equivalent because all conditions are satisfied. Hence, it follows that states 1 and 7 must be equivalent. States 2 and 6 would be equivalent states if it were not for the contradictory outputs.

In complex LSC's, the identification of equivalent states may prove to be a formidable task. To overcome this problem, a technique is presented which utilizes a chart called the Equivalent Pairs Chart (EPC). This chart consists of cells which uniquely relate two different machine states. A convenient form of the EPC is depicted by the lower diagonal matrix shown in Figure 7-3. An individual chart is provided for each column of the table, and each cell of the chart describes the equivalency or status of two stable states of the particular column. For example, in the EPC of Figure 7-3, the cell location N-M describes the equivalency of states N and M. Complying with the conditions establishing the equivalency of stable states, the EPC is completed by applying the steps given below:

Row No.	Outputs		Input States			
	Z_1	Z_2	I_1	I_2	I_3	I_4
1	0	0	[1]	2	–	3
2	0	1	1	[2]	4	5
3	1	1	–	–	4	[3]
4	1	0	7	6	[4]	3
5	–	1	–	6	4	[5]
6	1	1	7	[6]	4	–
7	0	0	[7]	2	–	5

Figure 7-2 An abridged LSC showing equivalent and nonequivalent states.

SEQUENTIAL LOGIC DESIGN—CLASSICAL SYNTHESIS

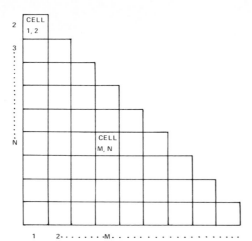

Figure 7-3 Equivalent pairs chart.

1. Insert an X in the appropriate cell if the outputs contradict each other.
2. Record in each cell the pairs of states which must be equivalent in order for the associated cell states to be equivalent.
3. Leave the cell blank if Steps **1** or **2** are not applicable.
4. Register the influence of a nonequivalent pair of states on each cell that contains the nonequivalent pair of states by placing an X in the respective cell. This step should be iteratively performed until all nonequivalent pairs of states have been considered.

After performing the steps above, the cells that are not X-ed indicate that the states defining the cell are equivalent to each other, while those involved with the X-ed cells are nonequivalent states.

In order to demonstrate the above procedure, consider the LSC shown in Figure 7-4. As there are only two stable states in the first and second columns of the LSC, the EPC for these columns consists of only one cell, as shown in Figure 7-5. For the first column, the equivalence of states 1 and 5 is dependent upon the equivalences of states 2 and 4, 6 and 10, and 3 and 13. Therefore, in accordance with Step **2** of the above procedure, the "cell" or cube is properly identified with the equivalency requirements: 2, 4; 6, 10; 3, 13. Likewise, for the second column, the single-celled EPC contains (1, 5; 9, 10; 3, 8) which are the three pairs of states that must be equivalent in order for states 2 and 4 to be equivalent. The EPC's for columns 3 and 4 are constructed in a similar manner. Note that at cell location (6, 9) an X is placed in the cube due to the contradictory outputs.

Once the EPC's are completed for each column, the next step is to check the equivalency of each pair of states. This is performed by taking an X-ed cube (one at a time) and recording its influence on all other cubes in each of the EPC's, whether they contain the coordinates of this X-ed cube or not. For example, consider the X-ed cells in Figure 7-5 for column 4 at location (7, 8), indicating the nonequivalence of states 7 and 8. Because of this nonequivalent condition, all other pairs of

190 INTRODUCTION TO FLUID LOGIC

Outputs		Input States			
A	B	I_1	I_2	I_3	I_4
0	0	〖1〗	2	6	3
0	1	5	〖2〗	10	8
–	1	1	–	12	〖3〗
0	1	1	〖4〗	9	3
0	0	〖5〗	4	10	13
1	1	–	4	〖6〗	3
0	1	5	–	14	〖7〗
1	1	1	–	15	〖8〗
1	0	–	2	〖9〗	3
1	–	–	2	〖10〗	7
–	1	–	–	〖11〗	3
1	–	5	4	〖12〗	3
0	1	1	–	6	〖13〗
1	–	–	–	〖14〗	8
1	0	–	4	〖15〗	16
0	–	5	2	11	〖16〗

Figure 7-4 An abridged logic specification chart.

states that depend upon the equivalency of these states for their equivalence are confirmed nonequivalent states. An X is hence entered in every cell location of the charts that contain this entry. For example, a (7, 8) is contained in cell location (10, 14); and therefore an X is entered in this location. Similarly, since states 10, 14 are now confirmed nonequivalent states, any states depending on them for their equivalence automatically becomes nonequivalent. The final EPC's for the problem of Figure 7-4 can be observed in Figure 7-5.

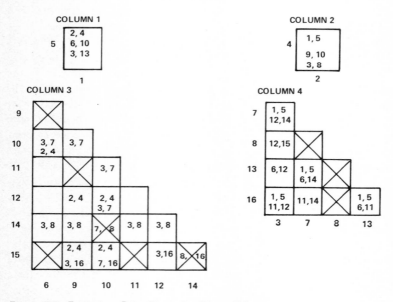

Figure 7-5 Equivalent Pairs Charts for Figure 7-4.

SEQUENTIAL LOGIC DESIGN—CLASSICAL SYNTHESIS 191

7-3 STATE SUBSTITUTION

Once the equivalency of the machine states is established, the states must be combined to produce an optimal specification chart containing the fewest number of new machine states or rows. This can be accomplished by a method called state substitution. The method employed here attempts to group the largest possible number of states into the fewest number of "new" states and by so doing achieves a near minimal state machine. The necessary conditions for grouping these states are:

1 Any two equivalent machine states can always be grouped into one set of states.
2 Three or more stable states can also be grouped to form a set of states if and only if each combination of two states satisfies the condition of equivalency.

The new set of states obtained is referred to as the "equivalent set" of machine states. A combination of equivalent states which conjoins the greatest possible number of stable states defines the "maximal equivalent set."

The equivalency status of the various states in the LSC was determined by the Equivalent Pairs Chart (EPC) in the previous section. However, in order to minimize the number of redundant states present in the LSC, it is necessary that the equivalent states be grouped such that they represent the greatest number of states. The group thus formed is the "maximal equivalent sets" of states that can serve to define the new state of the machine. As has been stated previously, such a set of states demands the equivalency of each pair of associated states. For small problems, the recognition of these "maximal equivalent sets" can be most conveniently achieved by using a graphical approach involving a nodal diagram as follows:

1 Initiate a nodal diagram for each EPC (column) by providing a node for each machine state represented in the EPC under consideration.
2 Connect each pair of nodes that are equivalent to each other as indicated by the EPC.
3 Group and identify the nodes that form the largest "complete" polygons. A complete polygon is one having not only its sides present but also all its diagonals. Note that a node can serve for more than one "maximal sized" polygon.

In the resulting nodal diagram (called the equivalency graph), the "maximal" polygons actually represent the maximal equivalent sets of the machine states; hence, all the states corresponding to the nodes of the polygon can be conjoined to form a single new machine state. In relation to the EPC for column 1 in Figure 7-5, it is easy to see that the equivalency graph involving states 1 and 5 is a two-node polygon (see Figure 7-6), and as this is a "complete" polygon, states 1 and 5 represents a "maximal equivalent set." Examination of the second EPC shows a similar situation between states 2 and 4.

The EPC for the third column is somewhat more complex. The formation of the nodal diagram can be observed in Figure 7-6. It is obvious that the search for

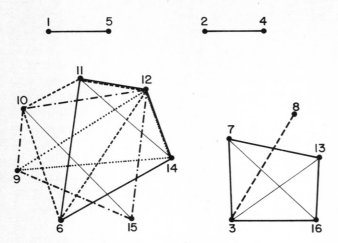

Figure 7-6 Equivalence graphs.

the maximal equivalent sets, or the largest complete polygons, in the graph, is not an easy task. However, since triangular forms in the graphs are easy to detect because they do not contain any diagonals, the following rationale should prove helpful for identifying the maximal polygons:

1 Locate a triangle in the graph.
2 Inspect all other nodes in the diagram not contained in the triangle and identify any node that is connected to each node of the triangle under consideration. The following cases may occur:
   a If no such connecting nodes exist, then the triangle is a maximal polygon, and its nodes represent a maximal equivalent set.
   b One connecting node exists which allows the formation of a four-node polygon and a maximal equivalent set involving four states.
   c Two or more nodes satisfy the interconnecting condition. In this case, every maximal complete polygon that can be formed by these nodes and the triangle under consideration constitutes a maximal polygon. When dealing with complex cases, the set of nodes formed in this step can be subjected to Steps 1 and 2 to find all maximal complete polygons. The term "all maximal complete polygons" implies that for each complete polygon, whether it is a single, double, trinode, or even multinode, there exists at least one maximal complete polygon that includes the earlier mentioned polygon.

To illustrate the identification process, consider the equivalency graph for the third column of Figure 7-6. Note that the existence of triangle (10, 11, 12) is easily detected. Inspecting other nodes reveals that only state 6 is equivalent to states 10, 11, 12, or that node 6 is connected to all nodes of the triangle (10, 11, 12). Therefore, the polygon (6, 10, 11, 12) is a maximal polygon. Furthermore, it can also be seen that nodes (9), (12), and (14) form a triangle; however, as no state is equivalent to all three states 9, 12, and 14, only these three states can be combined

SEQUENTIAL LOGIC DESIGN—CLASSICAL SYNTHESIS

to form a maximal equivalent set. The reader may verify that the maximal equivalent sets of this LSC is (1, 5); (2, 4); (6, 11, 12, 14); (6, 10, 11, 12); (9, 10, 12, 15); (9, 12, 14); (3, 8); and (3, 7, 13, 16).

Possessing the complete set of the "maximal equivalent sets" representing the machine states contained in the LSC allows the derivation of the necessary collection of equivalent sets. The following procedure can be applied to obtain the necessary sets to form a refined description of the original logic specification:

1. Form a machine state accounting table (See Table 7-1) as follows:
   a. List all states in the LSC at the top of the table.
   b. List all "maximal equivalent" sets on the left side of the table.
   c. Mark with an X every row and column location where the machine state of the column is included in the equivalent set that represents the particular row.
2. Bracket every X that occurs only once in a column and designate as essential (with an asterisk) those corresponding sets where one or more bracketed X's exist.
3. Place parentheses around all X's existing in a row containing a bracketed X.
4. Label the rows not containing bracketed or parenthetical X's by alphabetical characters, as illustrated in Table 7-1.
5. The essential sets are sufficient to represent all states of the LSC if a bracketed or parenthetical X is contained in every column of the machine state accounting table.
6. If the essential sets are insufficient, formulate a conjunctive logic expression which describes the combination of rows needed to cover each machine state (column) not satisfied by the essential sets.
7. Transform the conjunctive expression obtained in Step 6 into a disjunctive form and simplify.
8. The shortest product term in the simplified expression represents the minimal combination of sets (called the supplemental sets) needed to augment the essential sets.
9. Construct a Next-State Table by listing the essential and supplemental sets in the selected set column on the left side of the table.
10. For each selected set of states identify the corresponding "Next States" from the LSC and record them on the right side of the Next-State Table in groups having common input states (in the same column of the LSC).
11. Compare each next-state set with the selected sets and underline those next-state sets which are not included or covered completely by a single selected set.

Table 7-1 Machine State Accounting Table

	MAXIMAL EQUIVALENT SETS	STATES																
		1	2	3	4	5	6	7	8	9	10	11	12	13	14	15	16
*	(1, 5)	[X]				[X]											
*	(2, 4)		[X]		[X]												
A	(6, 11, 12, 14)						X					X	X		X		
B	(6, 10, 11, 12)						X				X	X	X				
*	(9, 10, 12, 15)									(X)	(X)		(X)			[X]	
C	(9, 12, 14)									X			X		X		
*	(3, 8)			(X)					[X]								
*	(3, 7, 13, 16)			(X)				[X]						[X]			[X]

Table 7-2 Next-State Table

	NO.	SELECTED SETS	NEXT STATE SETS
*	1	(1, 5)	(2, 4), <u>(6, 10)</u>, (3, 13)
	2	(2, 4)	(1, 5), (9, 10), (3, 8)
	3	(9, 10, 12, 15)	(5), (2, 4), (3, 7, 16)
	4	(3, 8)	(1), (12, 15)
	5	(3, 7, 13, 16)	(1, 5), (2), (6, 11, 12, 14)
	6	(6, 11, 12, 14)	(5), (4), (3, 8)
**	7	(6, 10, 11, 12)	(5), (2, 4), (3, 7)

*Underline (6, 10) per step **11**.
Auxiliary set added per step **12.

12 From the remaining maximal equivalent sets listed in the accounting table, select the least number of equivalent sets needed to cover all of the underlined next state sets and record these auxiliary sets together with their next state sets at the bottom of the Next-State Table.

13 Perform Steps **11** and **12** iteratively until all next-state coverage is obtained.

The procedure for identifying the necessary maximal equivalent sets can be illustrated by the tables shown in Tables 7-1 and 7-2. Note that all maximal equivalent sets together with state coverage information are presented in Table 7-1. Essential sets are indicated by an asterisk and the remaining sets labeled by A, B, and C. Eliminating all columns containing bracketed and parenthetical X's reveals that states 6, 11, and 14 are not covered by the essential sets. According to Steps **6**, **7**, and **8**, the required supplemental term can be derived as follows:

$$Z_c = (A + B)(A + B)(A + C) = A + BC \tag{7-1}$$

It can be noted that A is the simplest term that can satisfy the necessary coverage.

The essential and supplemental sets of states obtained from the accounting table need to be checked for closure; i.e., whether all next-state conditions are satisfied. This is accomplished in Table 7-2 by comparing the next state requirements with the selected sets derived from Table 7-1. Note that the next-state set (6, 10) is the only one not covered by the essential and supplemental sets. Therefore, it is necessary to utilize an auxiliary set from Table 7-1 to cover this next-state condition. This set is (6, 10, 11, 12) and it is recorded at the bottom of Table 7-2 together with its next states. If any of its next states were not covered by the selected sets, then other auxiliary sets would be required. Note that some of the selected sets representing the new machine states cover the original states more than once. This duplication is permitted as long as the conditions for state grouping are met.

7-4 REDUCED SPECIFICATION CHART

The final group of selected state sets identified in the Next-State Table represents the necessary states needed to describe the logic of the control network. These

SEQUENTIAL LOGIC DESIGN—CLASSICAL SYNTHESIS 195

states form the basis for a new description to replace the LSC—this is called the Reduced Specification Chart (RSC). To accommodate the construction of the RSC, each selected set is given a new identity by assigning arbitrary numbers to each set as shown in Table 7-2.

The format of the RSC is identical to that of the LSC. Entries in the chart are determined in accordance with the following procedure:

1. Place, in brackets, new state [1] in row 1 of column 1.
2. Identify the column with the input state associated with the original states of the selected set.
3. Label the row for the new state with a combined output state determined as follows:
   a. Compare the outputs of the selected set states on the LSC.
   b. Combine the outputs such that "don't care" outputs are absorbed by corresponding zeros or ones. "Don't care" entries are retained in the RSC if the output for each state is a "don't care" entry.
4. Place, in brackets, the next new state in the next row and in a new column if the input state is different than the ones in the previous columns. Otherwise, the next new state is placed in the column agreeing with its input state.
5. Repeat Steps 2 to 4 until all new states representing the selected sets are recorded.
6. Enter the new state numbers of the selected sets which cover each next state set in the columns where its corresponding stable (bracketed) state exists.

The RSC corresponding to the LSC in Figure 7-4 is shown in Figure 7-7. Note that a significant reduction in the number of rows has been achieved—16 rows in the LSC compared to seven rows in the RSC.

7-5 MINIMAL ROW SPECIFICATION CHART

After the redundant states have been eliminated from the logic specification, the resulting chart (RSC) contains the irreducible states of the machine. These states represent the actual states which the machine must achieve. Therefore, the state number given in the RSC must be preserved throughout the synthesis procedure.

OUTPUTS		INPUT STATES			
A	B	I_1	I_2	I_3	I_4
0	0	[1]	2	7	5
0	1	1	[2]	3	4
1	0	1	2	[3]	5
1	1	1	—	3	[4]
0	1	1	2	6	[5]
1	1	1	2	[6]	4
1	1	1	2	[7]	5

Figure 7-7 The Reduced Specification Chart.

One way to preserve the machine state numbers and yet reduce the size of the Logic Specification Chart is to assign more than one stable state per row. This row-minimization process can best be performed using a graphical approach similar to the one used for the redundant state elimination process discussed in the previous sections. Instead of searching for stable states which can be conjoined, rows are considered which contain compatible machine states. Two rows are recognized as compatible or mergeable if and only if there are no conflicting state numbers in any of the columns. In other words, two rows can be merged into one row if in each column there exist two state entries that are compatible—equal in value or where one contains an unspecified entry "–" in the next state location.

Because of the simplicity of the row-merging requirement, a compatibility graph can be constructed from the information in the RSC using the following rules:

1 Construct a nodal diagram such that each node corresponds to a necessary machine state (or row). Initially, these nodes or rows are disjointed from one another.
2 Check every two row combination for their mergeability. If they are mergeable, connect the two relating nodes by a line.
3 Group the nodes that form the largest complete polygon. Identify all possible maximal polygons which represent the maximal compatible sets of the RSC rows.
4 Utilize a row-accounting table similar to the one used in the previous section for reducing the number of rows of the LSC. This is accomplished by applying the following steps:
  a List all row state numbers at the top of the table.
  b List all maximal compatible sets on the left side of the table.
  c Mark with an X every row and column where the row or state numbers are included in the compatible set that represents the particular row.
  d Bracket every X that occurs only once in a column and designate as essential (with an asterisk) those corresponding sets where one or more bracketed X's exist.
  e Place parentheses around all X's existing in a row containing a bracketed X.
  f Label the rows not containing bracketed or parenthetical X's by alphabetical characters.
  g The essential sets are sufficient to represent all states of the RSC if a bracketed or parenthetical X is contained in every column of the machine-state accounting table.
  h If the essential sets are insufficient, formulate a conjunctive logic expression which describes the combination of rows needed to cover each machine state (column) not satisfied by the essential sets.
  i Transform the conjunctive expression obtained in Step 6 into a disjunctive form and simplify.
  j The shortest product term in the simplified expression represents the minimal combination of sets (called the supplemental sets) needed to augment the essential sets.
  k Remove any duplication of row representation from the essential or supplemental sets.

SEQUENTIAL LOGIC DESIGN—CLASSICAL SYNTHESIS   197

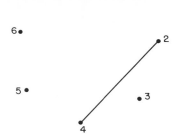

Figure 7-8 Compatibility graph for Figure 7-7.

For relatively small problems, it should be obvious that the important facets involved in Step **4** can be performed by the simple inspection of the nodal diagram. This is particularly true for the example problem discussed previously as shown in Figure 7-7. The row by row examination of this chart reveals that only two pairs of rows are compatible with each other and hence the compatibility graph shown in Figure 7-8 is quite trivial. Merging of rows 1 and 7 as well as rows 2 and 4 results in the Minimal Row Specification Chart (MSC) shown in Figure 7-9.

In order to demonstrate the above rules for a more extensive case, consider the RSC shown in Figure 7-10. The Compatibility Graph can be constructed as illustrated in Figure 7-11. By searching for the maximal polygons in the graph, a collection of maximal compatible sets is obtained, as listed in Table 7-3. Completing the table and applying the rules for state coverage presented in the previous section, it can be seen that three sets are actually mandatory for the representation, which are the sets (1, 3, 5), (2, 6, 9), and (4, 8, 11). Removing the rows represented by these sets, as well as the sets themselves, leaves ten sets and three rows undisturbed in the table. Labeling the remaining sets by the characters $A, B, \ldots J$, the logical expression for the supplemental coverage is as follows:

$$Z_c = (A + B + C)(A + C + H)(A + I + J)$$

Performing the distributive multiplication on the above expression yields the following disjunctive expression:

$$Z_c = A + BHI + BHJ + CI + CJ \qquad (7\text{-}2)$$

INPUT STATES			
I_1	I_2	I_3	I_4
(1)	2	(7)	5
1	(2)	3	(4)
1	2	(3)	5
1	2	6	(5)
1	2	(6)	4

Figure 7-9 The Minimal Row Specification Chart.

198   INTRODUCTION TO FLUID LOGIC

OUTPUTS		INPUT STATES			
R	Q	I_1	I_2	I_3	I_4
0	0	(1)	2	—	3
0	1	6	(2)	4	—
0	0	—	—	5	(3)
1	0	—	8	(4)	7
0	1	—	2	(5)	—
1	0	(6)	—	—	9
1	1	—	—	10	(7)
0	0	11	(8)	—	—
0	1	—	—	4	(9)
1	1	—	12	(10)	—
1	1	(11)	—	—	7
1	1	1	(12)	—	—

Figure 7-10 The Reduced Specification Chart.

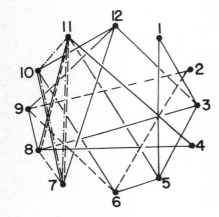

Figure 7-11 Compatibility Graph for Figure 7-10.

Table 7-3 Accounting Table

	MAXIMAL COMPATIBLE SETS	ROWS											
		1	2	3	4	5	6	7	8	9	10	11	12
*	(1, 3, 5)	[X]		X		X							
*	(2, 6, 9)		[X]				X			X			
*	(4, 8, 11)			[X]					X			X	
A	(7, 10, 12)							X			X		X
B	(7, 8, 11)							X	X			X	
C	(7, 10, 11)							X			X	X	
D	(8, 9)								X	X			
E	(3, 8)			X					X				
F	(5, 6)					X	X						
G	(5, 11)					X						X	
H	(6, 10)						X				X		
I	(3, 12)			X									X
J	(9, 12)									X			X

SEQUENTIAL LOGIC DESIGN—CLASSICAL SYNTHESIS

I_1	I_2	I_3	I_4
(1)	2	(5)	(3)
(6)	(2)	4	(9)
(11)	(8)	(4)	7
1	(12)	(10)	(7)

Figure 7-12 The Minimal Specification Chart (MSC) for Figure 7-10.

In this expression, A is the shortest term in the expression and therefore the supplemental set (7, 10, 12) is selected. The resulting four-row MSC is illustrated in Figure 7-12.

Since more than one stable state exists per row, it is important to note that the recording of the outputs cannot be accomplished as in previous cases. To transfer output information requires the expansion of the output subtable. This will be illustrated in the following sections by revealing that the output subtable does not influence the configuration of the specification table except for the generation of the outputs themselves. Hence the output information is excluded from the MSC (Figure 7-9 and Figure 7-12).

7-6 OPERATIONAL FLOW CHART

The Minimal Row Specification Chart (MSC) is the most compact form of the desired network specification. Since each row of the MSC relates to a particular memory state, it is necessary that the rows be properly labeled in binary (GRAY) code form in order to facilitate the construction of the network Karnaugh maps. The order or arrangement of the rows of the MSC has been made in a purely arbitrary manner without any consideration for the transitions involved in changing from one machine state to the other. In the course of ordering the rows to form the Operational Flow Chart (OFC), it may prove necessary to increase the number of rows of the specification.

Like the Karnaugh map, transitions between adjacent rows and columns are acceptable because they require only single changes to be made at a time in one of the map parameters. Where a transition requires row or column "jumping" to occur in order to reach the "next state," a condition called a race takes place which can make the logic operation indeterminant. Such transitions must be avoided at all costs and therefore their features should become familiar enough to the reader that he can detect their existence by mere inspection.

In order to illustrate an OFC which exhibits critical transitions, consider the one shown in Figure 7-13. A two-memory variable Gray Code was assigned to

Y_1 Y_2	I_1	I_2	I_3	I_4
0 0	(1)	2	(5)	(3)
0 1	(11)	(8)	(4)	7
1 1	(6)	(2)	4	(9)
1 0	1	(12)	4	(7)

Figure 7-13 Memory transitions in a flow chart.

distinguish the four rows of the MSC from one another. Note the situation which will occur when the machine is at state 1 and the input changes to I_2, necessitating a transfer to the machine state 2. Since unstable state 2 has a memory state ($Y_1 Y_2 = 00$), while its associated stable state 2 is ($Y_1 Y_2 = 11$), a simultaneous activation of both memory elements is needed to accomplish the desired operation—a situation which might be fatal if not corrected. With one of the memory elements being faster than the other, the control system will "suddenly" find itself in stable state 8 or 12 and the memory transition process is terminated in an undesired state. This same situation can occur with the transition associated with unstable state 7 when transferring to stable state 7.

In order to avoid this "race" phenomenon, memory-state assignments must be made such that only single memory changes occur at a time. To accomplish this predictable operating condition, rows may be added to the MSC even though it increases the number of memory elements required for representing the system. Of course, as with any binary number system, it takes N variables to uniquely represent 2^N entities (in this case rows). For the MSC shown in Figure 7-13, correct transitions can be achieved if another memory element is added to the control network, thus creating an eight-row chart as shown in Figure 7-14. Note that the transitions are "forced" to transfer in four steps. Since the chart in Figure 7-14 satisfies all the required transitions of the memory states, it is referred to as the Operational Flow Chart (OFC). More specifically, an OFC is one where each of its rows corresponds to a memory state of the system and where only single memory transitions are permitted.

Although the row-addition approach can always be used to achieve a valid OFC, it is not likely that a minimal hardware network can be obtained when such an approach is utilized. It is therefore the primary objective of this section to try to satisfy all memory transitions in a minimal-row OFC without resorting to row additions. It is maintained that a minimal memory circuit resulting from such an OFC is the best approximation to an absolute minimal network. The following conditions and definitions are needed for the development of the minimal row OFC:

Condition 1 There are always 2^N rows in the OFC, where N is an integer such that 2^N is at least equal to the number of rows in the related MSC. Basically, this condition means that the number of rows of the OFC cannot be less than the number of rows in the MSC. Furthermore, the number of rows in the OFC is

Y_1 Y_2 Y_3	I_1	I_2	I_3	I_4
0 0 0	(1)	2	(5)	(3)
0 0 1	(11)	(8)	(4)	7
0 1 1	(6)	(2)	4	(9)
0 1 0	1	(12)	4	(7)
1 1 0	—	—	—	7
1 1 1	—	2	—	7
1 0 1	—	2	—	7
1 0 0	—	2	—	—

Figure 7-14 Eight-row Operational Flow Chart.

SEQUENTIAL LOGIC DESIGN—CLASSICAL SYNTHESIS

defined by the number of memory states available, i.e., 2^N rows when N memory elements are utilized.

Definition 1 *Adjacent rows*: Two rows are adjacent to each other if there is only one memory-variable value change between the two memory states that represent the rows.

The adjacency criterion used here is the same as the one used for the Karnaugh maps—two cells are considered adjacent if they have exactly one variable value difference between the representing terms. In relation to the OFC of Figure 7-14, it can be noted that row 1 is adjacent to rows 2, 4, and 8; row 2 is adjacent to rows 1, 3, and 7, and so forth.

Condition 2 Each row of a 2^N row OFC has exactly N adjacent rows. To illustrate this condition, consider the three-memory example shown in Figure 7-14. It can be seen that the fourth row, represented by the memory state $(Y_1 Y_2 Y_3 = 010)$, has three adjacent rows: row 1, row 3, and row 5.

Definition 2 *Primary adjacency*: An adjacency condition that must be satisfied between two rows—one containing a stable state and the other containing its only corresponding unstable state.

For example, the adjacency requirement between rows 1 and 4 of Figure 7-13 is a primary adjacency due to state 1. Note that there is only one unstable state 1 represented in the table.

Definition 3 *Secondary adjacency*: An adjacency condition that must be satisfied between a row containing a stable state and two or more rows containing corresponding unstable states.

In Figure 7-13, the second, third, and fourth rows have a secondary adjacency requirement, due to machine state 4. At this stage, it is important to note that a primary adjacency can be transformed into a secondary adjacency by assigning an unstable state of the state under consideration to one or more unspecified locations. For example, in Figure 7-14, the primary adjacency requirement between rows 1 and 3 (due to state 2), has been transformed to a secondary adjacency requirement between rows 1, 3, 6, 7, and 8.

The problem of formulating the OFC is one of satisfying all primary and secondary adjacencies and yet retaining the minimal-row flow chart. There have been no pure mathematical means to formulate the minimal-row configuration of an OFC, not to mention the "actual minimal" form of the OFC. The approach used in this book involves a graphical means for revealing the near-minimal form of the OFC. The following steps are required:

1 Determine the least possible number of rows of the OFC with respect to the MSC. Particularly, select the smallest value for N such that 2^N is at least equal to the number of rows in the MSC. Place the rows of the MSC in the upper part of the "unarranged" flow chart.

2 Develop an adjacency graph for each row of the flow chart. This graph consists of a principal node representing a given row and secondary nodes representing rows having adjacency requirements. Secondary nodes will have single-row designations to indicate primary adjacencies. Join each secondary node with the primary node and indicate on the associated branch the state requiring the adjacency (see Figure 7-15).

3 For each adjacency graph, merge the branches that have identical secondary nodes and record the state numbers on the conjoined branch.

4 For graphs which contain more than N (number of memory) branches, grouping must be performed as follows:
 a Classify all primary adjacent rows as mandatory.
 b Signify these mandatory rows by circling the row designation at all secondary nodes.
 c Count the number of branches that do not contain any circles and add the number of different mandatory adjacent rows. If the total does not exceed N, proceed to the next graph.
 d If N is exceeded, select a row that occurs the most (at least twice) in all branches not having circled rows. Classify this row as mandatory adjacent and return to Step 4 b.

When Step 4 c is satisfied for all existing graphs, proceed to Step 7.

5 In each adjacency graph where Step 4 is not satisfied, record at the secondary node the row designation where a dash exists in the column of the associated state (put row label in parentheses). When appropriate dashes do not exist for the critical graphs, the size of the flow chart must be doubled (increment N) and return to Step 2.

6 In the adjacency graphs considered in Step 5, select a parenthetical entry that exists with a prime adjacency. This entry must also exist at one of the secondary nodes outside a parenthesis. In addition, the selected entry must satisfy at least one of the following hierarchically arranged conditions:
 a It is mandatory adjacent to the principal row.
 b It has a mandatory adjacency relationship with a row outside the parentheses under consideration.
 c It has a secondary adjacency relationship with one of the rows outside the parentheses under consideration. If such cases do not exist, check whether or not there are parenthetical entries that occur in two different branches. If

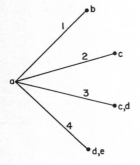

Figure 7-15 An adjacency graph.

more than one entry satisfies this condition, select the pair of branches which have state numbers that represent the most branches in the adjacency graphs considered in Step 5. It should also be noted that all rows involved in an adjacency requirement must be interconnected. Therefore, it is important that the selection of a parenthetical entry is made such that it will not create new adjacency requirements whenever possible. The entry that satisfies these conditions is placed outside the parentheses in the associated branch(es). Using the approach in Step 4, check whether or not this step reduces the mandatory adjacencies of the graph at least by one. When these conditions are satisfied, reorganize the flow chart by inserting the identified state as indicated by the associated branch and reorganize the graphs. If not satisfied, place the entry back into the parentheses and select another row, where possible. If such a case does not exist, increment N, double the size of the flow chart, and return to Step 2.

7 Construct a Karnaugh map using the N memories as the Karnaugh variables.
8 Select the adjacency graph that contains the most mandatory adjacencies and place the principal row in the "all zero" position of the Karnaugh map.
9 At adjacent locations, place the associated mandatory adjacent rows in an arbitrary manner. The secondary adjacencies are satisfied using the following rules:

   a If the secondary adjacency requirement involves a mandatory adjacency, following the assignment of the mandatory adjacent row to a cell, all other unassigned rows involved in the adjacency requirement are entered in parentheses in all unassigned locations that are adjacent to the particular cell. A cube or cell is said to be assigned to a particular row if the cell will definitely be utilized for the row under consideration. These parenthetical entries indicate that the particular cube has become a candidate location for the particular entries.
   b If the secondary adjacency requirement does not involve a mandatory adjacency, insert all rows involved in the adjacency requirement in an adjacent cube and all unassigned locations adjacent to the particular cube (in parentheses).

10 In all unassigned cubes of the adjacency Karnaugh map, enter the unassigned mandatory adjacent rows in accordance with their adjacency requirements specified by the adjacency graphs.
11 Search the map for a row that is needed the most in a particular position as indicated by the number of times such a row is represented in the particular cell. Assign this row to the particular cell unless it jeopardizes the fulfillment of a secondary adjacency requirement. An assignment of a row to a location that is needed to satisfy an adjacency requirement can only be made if either the row is involved in the adjacency requirement or if the row possesses a "don't care" that can be utilized to satisfy the particular adjacency.
12 Inspect the map for a row that has only one possible location in the map. Assign the critical row to the particular position.
13 Perform Steps 10, 11, 12 iteratively until no other possibilities exist.
14 If unassigned rows still exist, assign them to the remaining locations in order to complete the adjacency requirements of the system. Double the size of the flow chart if the adjacency requirements are still not satisfied.

	I_1	I_2	I_3	I_4
a	(1)	2	(3)	(4)
b	7	(2)	5	(6)
c	1	(8)	5	—
d	(7)	8	3	(9)
e	(10)	8	(5)	4
f	7	(11)	12	9
g	(13)	11	—	6
h	(15)	11	(12)	(16)

Figure 7-16 Minimal Specification Chart.

In order to provide the reader with a working knowledge of the row-arranging procedure, an illustrative example is presented in Figure 7-16. As there are eight rows in the MSC, the value of N is three, and therefore, to obtain a minimal-row OFC, the first trial should consider an eight-row OFC.

The adjacency graphs (Figure 7-17) are developed directly from the adjacency relationships of the rows of the MSC. For example, observe that row a is associated with row c because of Machine State 1; therefore, a branch is constructed to indicate this adjacency connection. This branch is shown in both adjacency graphs of rows a and c. Similarly, row a has a primary adjacency with row b due to state 2, and so on.

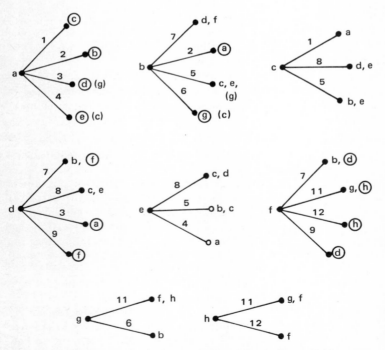

Figure 7-17 Adjacency graphs for example problem in Figure 7-16.

After completing all adjacency graphs, it can be noted that there are four branches emanating from the adjacency graphs of rows a, b, d, and f. As this number exceeds the value of N (= 3), a process of merging has to be performed, if possible. In the case of row d, row f is mandatory adjacent, and a circle is made around the row f designation to indicate mandatory adjacency of row f and row d. This mandatory adjacency requirement is transformed to the adjacency graph of row f, where the secondary adjacency (f-b,d) is merged to the primary adjacency (f-d). (Circle the row f designation.)

A different situation exists for the adjacency graphs of rows a and b. In these cases, no merging can be performed, as none of the conditions of Step **4** can be satisfied. Therefore, in accordance with Step **5**, the "don't care" locations are utilized in these two graphs. For example, since there is a "don't care" cell available in column 3 at row g of the MSC that might be useful to state 3, a g is placed in parentheses in the branch associated with state 3, as shown in Figure 7-17.

Having completed the adjacency graphs, Step **6** can be performed. Arbitrarily, consider the first adjacency graph relating to row a. In this graph, note that row g is not represented anywhere on the graph, which leaves no choice but to select row c, which has a primary adjacency to row a due to state 1. Hence, an unstable state—4—is inserted in the "don't care" position of row c of the MSC, which results in the modifications of the adjacency graphs (see Figure 7-18(*a*), (*b*), (*c*), and (*d*)). Similarly the transformation of the adjacency graph of row b can be accomplished (see Figure 7-18(*e*)).

The next step is to construct the adjacency Karnaugh map for an eight-row flow chart. The map is coded to accommodate the three memory variables in a Gray Code format, as illustrated in Figure 7-19. For the 000 position of the map, select the adjacency graph that has the most mandatory adjacencies; in this case, select row a, as it has three mandatory adjacencies with rows b, c, and d. With row a placed at the 000 cell, place rows b, c, and d at the adjacent map locations. Because of the secondary adjacency of row a with rows e and c, either row a or row c has to be adjacent to row e; hence, at this stage, the location of row e has two alternatives—to be placed at cube location 011 or at cube 110 of the map. As these are only tentative locations, e is inserted in parentheses, as shown in Figure 7-19.

As the locations for rows b, c, and d are finalized, their adjacency graphs must be considered. For row b, there is a mandatory adjacency with row g, which can only be satisfied by two positions, i.e., node locations 101 and 011. Again, g is placed in parentheses at both locations. Similarly, inspection of the adjacency graph of row d reveals that row f must be placed in locations 101 and 110. Note that row c does not have a mandatory adjacency except row a, which has been assigned to location 000. At this point, the following conclusions can be drawn:

- At locations 011, 110, and 101 rows e, f, and g are all required exactly once, and hence, no conclusion can be drawn using this basis.
- Observing the adjacency graph of row c, it is evident that rows c and d are involved in a secondary adjacency due to state 8. Inserting row f in location 110

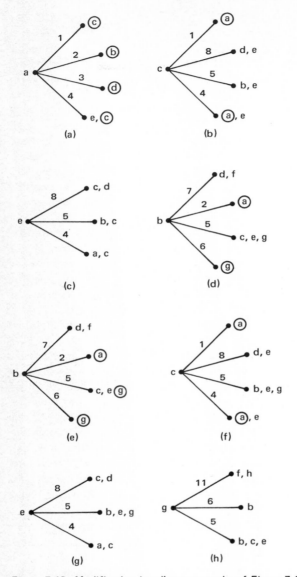

Figure 7-18 Modification in adjacency graphs of Figure 7-17.

would definitely jeopardize the possibility of satisfying this adjacency requirement, while the insertion of row e in this location satisfies this condition as e is included in the adjacency connection.
- Similarly, row f is required by a secondary adjacency with rows b and d, which shows that the selection of location 101 for row f is justifiable. Note that if, for example, row f is placed in location 111, any row that occupies locations 110 or 101 should have "-" entries in appropriate columns (in this case column 1 and

SEQUENTIAL LOGIC DESIGN—CLASSICAL SYNTHESIS

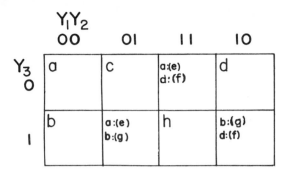

Figure 7-19 Adjacency Karnaugh map.

4) in order to allow the fulfillment of the primary and secondary adjacency with rows b and d.
- This leaves no choice other than assigning row g to location 011, due to the primary adjacency requirement to row b.
- The adjacency map is completed by assigning row h to location 111.

Reviewing the adjacency graphs of each row, it can be seen that the completed adjacency map satisfies all the adjacency requirements of the MSC. The resulting OFC can be observed in Figure 7-20. At this stage it may be important to note that there is only one solution for this problem without doubling the size of the OFC.

As powerful as it may seem, the method presented here may pose unnecessary difficulties when the problem is quite simple. For these cases, inspection and intuition may suffice for obtaining the minimal-row OFC. The reader is again reminded that the actual minimal hardware network may not always be obtained with a minimal-row OFC.

7-7 EXCITATION CHARTS AND MAPS

The OFC provides the unique state descriptions needed for the synthesis of the control network. The benefits derived from the effort expended in reducing the number of rows and arranging them to eliminate critical transitions will become increasingly apparent. It has been mentioned earlier that each row of the flow chart is related to an internal, memory, or secondary state of the machine. This is especially true for the Operational Flow Chart, since each row was assigned to a

Y_1	Y_2	Y_3	I_1	I_2	I_3	I_4
0	0	0	(1)	2	(3)	(4)
0	0	1	7	(2)	5	(6)
0	1	1	(13)	11	5	6
0	1	0	1	(8)	5	4
1	1	0	(10)	8	(5)	4
1	1	1	(15)	11	12	(16)
1	0	1	7	(11)	(12)	9
1	0	0	(7)	8	3	(9)

Figure 7-20 Operational Flow Chart for the MSC of Figure 7-16.

particular memory state, as illustrated in Figure 7-20. The OFC was formulated carefully so as to permit only one memory variable to change for each transition. Thus, the relationship of each memory element to the rest of the system was established by the construction of the OFC. In order to develop the excitation maps for a specific problem, its OFC must be transformed such that each machine state in the chart is expressed in binary form.

The transformation of the machine states into binary form produces what is called an "Excitation Chart." This transformation can be performed by direct inspection of the OFC by simply substituting the memory state identifying the OFC row for all stable states in the row. The binary designation for an unstable state will be the binary label of the OFC row which represents the next transitional state. Thus, when a machine state agrees with its memory state, it is stable and a change in the input is required to cause the machine state to change. But when the machine state does not agree with the memory state, the memory is forced to change its state by one variable as dictated by the binary value of the machine state. If the new machine state is not stable (agree with its memory state), then the memory state is forced to change again—always in reach for a stable machine state.

To illustrate the transformation of an OFC into an Excitation Chart, consider the OFC shown in Figure 7-20 and the resulting excitation chart presented in Figure 7-21. In the first row of the OFC of Figure 7-20, it can be observed that states 1, 3, and 4 are stable states. The memory state for this particular row is (000), and therefore (000)'s are entered in the corresponding map locations in Figure 7-21. Similarly, in the second row, the memory state (001) is entered at stable state locations for states 2 and 6.

The memory or secondary-state assignment for unstable machine states needs more careful consideration. Essentially, the assignment should always force the memory state of the machine to change such that the next desired machine state is reached. To accomplish this, the next desired memory state is inserted in the chart location shown by the "001" entry in the first row of Figure 7-21. When two or more unstable states are present (as in the case of state 5), the memory-state assignment should reflect the "nearest" next state in which the next single-memory transition reaches. In the second row of Figure 7-21, the associated memory state is (001), while stable state 5 is appropriately identified in the fifth row by the

$Y_1 Y_2 Y_3$	I_1	I_2	I_3	I_4
000	(000)	001	(000)	(000)
001	101	(001)	011	(001)
011	(011)	111	010	001
010	000	(010)	110	000
110	(110)	010	(110)	010
111	(111)	101	101	(111)
101	100	(101)	(101)	100
100	(100)	110	000	(100)

Figure 7-21 The Excitation Chart.

SEQUENTIAL LOGIC DESIGN—CLASSICAL SYNTHESIS

memory state (110). Three memory transitions are required to properly transcend from (001) to (110), since it could be disastrous for "110" to be assigned to row 2 in the location of unstable state 5. Figure 7-20 illustrates the appropriate way of transcending to stable state 5; i.e., transition in stages to the third, then to the fourth, and finally to the fifth row, where stable state 5 exists. This type of transitioning can be reflected on the excitation chart by inserting in the second row a machine-state value of "011," which forces the activation of memory Y_2 and the subsequent switching of the network to the third row of the chart. Assigning this third-row location with the secondary state "010" causes the system to remain unstable and to initiate a similar transition to the fourth row. Finally, the transition to the fifth row, where the associated stable state exists, is accomplished by entering "110" in row 4.

Once the Excitation Chart is completed, a Karnaugh map can be formed and the logic expression for the secondary circuits of the network can be derived by applying the rules associated with Karnaugh maps. Note that each machine state in the excitation chart agrees to a corresponding cell location in the Karnaugh map. Therefore, a transposition of each state in the excitation map to a cell in the Karnaugh map can be achieved by considering unassigned positions as "don't cares." For example, if the input states associated with the excitation chart of Figure 7-21 are 00, 01, 11, 10, corresponding to I_1, I_2, I_3, and I_4, respectively, then the Karnaugh excitation map representation for the secondary or memory function of the network can be expressed as in Figure 7-22. The associated secondary equations can be written as follows:

$$\begin{aligned}
Y_1 &= A'B'Y_2'Y_3 + A'BY_2Y_3 + ABY_2Y_3' + Y_1Y_3 + A'B'Y_1 + A'Y_1Y_2' \\
&\quad + B'Y_1Y_2' \\
Y_2 &= A'Y_1'Y_2Y_3 + ABY_1'Y_3 + BY_2Y_3' + B'Y_1Y_2 + A'BY_1Y_3' + BY_1'Y_2 \\
&\quad + Y_1Y_2Y_3' \\
Y_3 &= A'BY_1'Y_2' + A'Y_1'Y_3 + B'Y_2Y_3 + Y_1'Y_2'Y_3 + BY_1Y_3 + Y_1Y_2Y_3 \\
&\quad + BY_2'Y_3 + A'Y_2Y_3 + B'Y_1'Y_3
\end{aligned} \qquad (7\text{-}3)$$

The above expressions are the static-hazard-free representations of the memory functions as defined and discussed in Section 6-9. The reader may verify that the actual minimal form of the equations are represented by omitting the last two terms of the Y_2 representation and the last four terms of the Y_3 representation. When these equations are to be implemented with hardware using logic elements other than memory elements, the hazard-free representation must be utilized; otherwise the deviation in the output of a memory element may cause a permanent unwanted output. For example, consider the case of the third memory equation where the hazard-eliminating term $Y_1Y_2Y_3$ is excluded from the expression. If Y_1, Y_2 and Y_3 are all activated, while initially B is ON, then the memory element Y_3 will remain in its ON condition due to the term BY_1Y_3. Furthermore, if the input B is deactivated, the function indicates that Y_3 should remain activated, since it is represented by the term $B'Y_2Y_3$. Thus, a delay can occur during the activation of

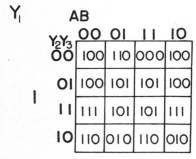

Figure 7-22 Karnaugh excitation maps of the secondary equations.

this latter term which, if Y_3 has been deactivated in the meantime, could result in the permanent deactivation of Y_3.

In the case where memory elements are utilized, it is important to identify the appropriate SET and RESET equations needed to operate each of the memory functions. Obviously, since a minimal representation is desired, these SET and RESET equations should be obtained from the Karnaugh excitation maps associated with the memory functions. Also, since the function described by the Karnaugh map is actually the excitation function and the map variables are the current outputs of the memories, an entry in the map should be compared with the condition or state of the particular memory and utilized to establish the necessary SET or RESET states of the particular memory device. Specifically, if the condition or state of the memory is 0, while the entry in the map is 1, then this cube represents a machine state where the activation of the particular memory is mandatory. In this case, an S is inserted in this particular location, thus replacing the existing 1 in the cube. Similarly, an R is placed in the map location where a 0 is found and where the memory condition or state is 1. To illustrate this SET/RESET assignment technique, consider the Karnaugh map for the Y_1 memory function shown in Figure 7-23. In this map, 1's are found in map locations ($ABY_1Y_2Y_3 =$ 00001, 01011, and 11010). Note that at these three locations the condition for Y_1 is 0; consequently, these are mandatory SET locations for Y_1, and thus S's are placed in the corresponding map locations. Similarly, R's are placed in the locations where 0's exist in the right side Karnaugh map in Figure 7-23.

From the reconstructed Karnaugh map, the minimal representation of the memory functions can be obtained. The characteristics of the memory play an

SEQUENTIAL LOGIC DESIGN—CLASSICAL SYNTHESIS

important part in establishing the degree of minimization which can be achieved by the function. With a few exceptions, the SET equations needed to properly control all commercially available memory devices (those exhibiting SET and RESET features) can be simplified by assuming the S's as the 1's of the SET equation and all the 1's and the "don't cares" of the memory function as "don't cares." Similarly, the RESET equation can be effectively simplified by taking the R's as the 1's while considering all 0's and "don't cares" of the memory function as the "don't cares" of the RESET expression. The resulting simplified Karnaugh maps for the SET and RESET equations of the Y_1 memory function are shown in Figure 7-24(a,b). The SET and RESET equations can be written as follows:

$$\text{SET}_{Y_1} = A'B'Y_2'Y_3 + A'BY_2Y_3 + ABY_2Y_3' \tag{7-4}$$
$$\text{RESET}_{Y_1} = A'BY_2Y_3' + ABY_2'Y_3' + AB'Y_2Y_3' \tag{7-5}$$

Special Memory Implementation

To illustrate the interpretation of the SET and RESET characteristics of the memory devices, the Karnaugh map shown in Figure 7-25 is considered. This map represents the memory function Y_1. Applying the rules for deriving the SET and RESET equations, the following expressions are obtained:

$$\text{SET}_{Y_1} = B' \tag{7-6}$$
$$\text{RESET}_{Y_1} = A'BY_2' \tag{7-7}$$

The above equations are, of course, applicable for all types of memory elements exhibiting SET and RESET features. However, consider the situation where a FLUID MEMORY-type moving part device as discussed in Chapter 4 is used for implementing the memory function. The truth table for this FLUID MEMORY device is shown in Table 7-4. In this table, it can be observed that the particular memory element exhibits the condition where it is activated whenever the SET signal is energized, no matter what the condition of the RESET signal. The

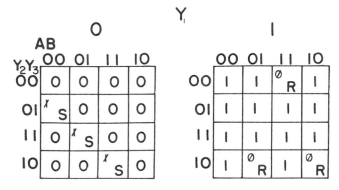

Figure 7-23 Reconstructed Karnaugh map representation for Y_1.

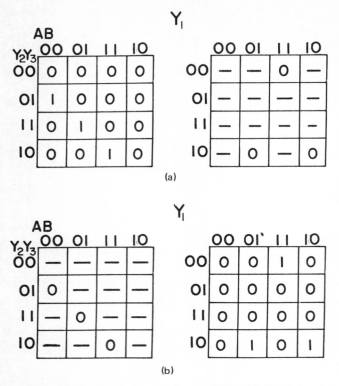

Figure 7-24 The Karnaugh maps for the SET and RESET equations: (a) SET$_{Y_1}$; RESET$_{Y_1}$.

development of the SET equation is the same as the above—i.e., no 0's are included in the SET representation, as this results in the activation of the particular memory element. On the other hand, theoretically, all map positions that have been used for the SET function can be used as a "don't care" for the RESET function. However, the use of "don't cares" may result in complications where the SET and RESET equations are deactivated at the same time. Of course, the one that is deactivated the last (due to a longer delay) will be prevalent, and an indeterminate output condition results. In order to prevent this, the following rules may be applied:

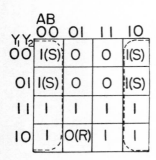

Figure 7-25 Karnaugh map for the MEMORY function of Y_1.

SEQUENTIAL LOGIC DESIGN—CLASSICAL SYNTHESIS

Table 7-4 The Truth Table for the FLUID MEMORY Element

S	R	Y*	Y
0	0	0	0
0	0	1	1
0	1	1	0
0	1	0	0
1	1	0	1
1	1	1	1
1	0	1	1
1	0	0	1

Y* = Previous condition of Y.

1. All S's are made mandatory SET states.
2. All 1's and "don't cares" of the memory function are considered as the "don't cares" of the SET equation.
3. Once the minimum SET function has been selected, the Karnaugh map of the memory function should be searched for 1's that are included in the representation of the SET function. Determine whether a 1 exists inside the SET function which differs from a 1 outside the SET function by only one change in the inputs. Select the 1's inside the SET function that do not exhibit the above adjacency condition to be the "don't cares" of the RESET function.
4. All R's are considered mandatory RESET states.
5. All 0, "don't care," and 1 locations resulting from Step 3 should be considered as the "don't cares" of the RESET function.

Using these rules, the simultaneous deactivation of the SET and RESET equations can be avoided. Note that in the Karnaugh map shown in Figure 7-25 where the SET term has been determined as in Equation 7-6, only five map locations satisfy the condition of Step 3, which are $ABY_1Y_2 = (0000,0001,1000,1001,0010)$. Therefore, the Karnaugh map for the RESET function can be constructed as in Figure 7-26 and described as follows:

$$\text{RESET}_{Y_1} = A'Y_2' \tag{7-8}$$

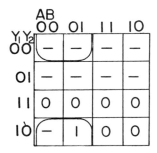

Figure 7-26 RESET function for the fluid memory implementation of Y_1.

This expression for the RESET condition is a more simplified solution than Equation 7-7.

A more complex situation can occur when it is necessary to obtain a maximal simplification by utilizing a detented memory valve. For example, consider the situation where the function of one of the memory elements is shown in Figure 7-27. The sixteen machine states are numbered in order to help the reader follow the discussion. Assume that states 6, 7, and 11 are identified as the unstable states of the machine, and if only single variable changes are permitted, state 6 can only occur after state 5 as well as state 7 after state 8. Assume also that the machine is "somewhat" deterministic—i.e., state 11 cannot occur after state 12. By using the "standard" means of deriving the SET and RESET function of this memory element (Y_1), the following expressions are obtained:

$$\text{SET}_{Y_1} = A'BY_2 \tag{7-9}$$
$$\text{RESET}_{Y_1} = AY_2' + B'Y_2 \tag{7-10}$$

Note, however, that if the truth table of the detented memory is evaluated, the memory will remain stable in its previous condition when opposing SET and RESET signals are exposed to the memory valve (see Table 7-5). This stability condition, of course, will depend upon the value of the detent force of the valve. Utilizing the unique characteristics of memory elements, the following procedure may be applied to achieve optimal implementation:

1. Inspect the LSC to identify the states that always follow another state. (Note that unstable states always follow a stable state, which in turn leads to its corresponding stable state.) Record such information on the Karnaugh map.
2. Label each 1 and 0 entry by an X if they follow a state having 1 and 0, respectively (with no change).
3. Determine the S and R locations.
4. All S's are the mandatory SET states.
5. All 1's and "don't cares" are used as the "don't cares" for the SET function.
6. All R's are the mandatory RESET states.
7. All 0's and "don't cares" are used as the "don't cares" for the RESET function.
8. Try to utilize X-ed 1 entries as "don't cares" for the RESET function and also try to simplify the SET function by using the X-ed 0 entries. As soon as an X-ed

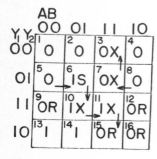

Figure 7-27 Karnaugh map for the memory function of Y_1.

SEQUENTIAL LOGIC DESIGN—CLASSICAL SYNTHESIS

Table 7-5 The Truth Table for the FLIP-FLOP MEMORY Element with Detent

S	R	Y*	Y
0	0	0	0
0	0	1	1
0	1	1	0
0	1	0	0
1	1	0	0
1	1	1	1
1	0	1	1
1	0	0	1

Y* = Previous condition of Y.

entry is included in a RESET function, it becomes a mandatory SET state, and vice versa. Determine whether a simpler SET and RESET function results.

In order to illustrate this procedure, consider the Karnaugh map shown in Figure 7-27. The "flow" as represented in the LSC is projected by the arrows to show the state changes in the map. X's are placed in state locations 3, 7, 10, and 11, showing that there are no changes in the state of the memory during the particular transition. Having determined the positions of the mandatory SET and RESET, it can be seen that state 7 can be used as a "don't care" for the SET function and that state 11 can be used as the "don't care" for the RESET function. This requires that they be included in both SET and RESET functions, as indicated by Step **8**. The resulting equations are:

$$\text{SET}_{Y_1} = BY_2 \tag{7-11}$$
$$\text{RESET}_{Y_1} = A + B'Y_2 \tag{7-12}$$

These equations are obviously more simplified than the previous standard results of Equations 7-9 and 7-10.

The reader may already note at this point that all the above "unstandard" means of obtaining the SET and RESET functions consider the possibility of the simultaneous actuation of both sides of the memory valve. However, other types of memory devices may exhibit different characteristics such that the formulations of a procedure similar to the one given above may be necessary. The reader should note that the "standard" means is valid for any type of memory device exhibiting the SET and RESET feature.

7-8 OUTPUT MAPS

The classical synthesis method ignores the outputs of the system after the Reduced Specification Chart is constructed. However, in order to derive the output equations for the network, the outputs must be recorded appropriately on the OFC. As with the excitation maps the output map consists of rows and columns—each row designates a secondary state, while each column reflects an input state. Entries in the map are made using the following rules:

1. Enter the output state associated with each stable state of the RSC in the location corresponding to the particular stable state in the OFC.
2. If two stable states are involved in a transition, then the outputs of all unstable states involved in the transition must be either:
   a. The same, if both stable states exhibit the same state of the particular output.
   b. A "don't care," if otherwise.

Actually, only Rule 1 is a necessary condition for the correct representation of the system. However, in order to avoid transient oscillatory changes in the outputs due to a state change, both of the above rules should be satisfied. The completed output map describes the conditions under which the prescribed outputs will occur and, therefore, it can be directly related to the Karnaugh map in the same way as that accomplished for the excitation map.

To illustrate the use of the above rules, the example of the previous section is discussed. The Minimal Specification Chart was shown in Figure 7-12 and is repeated in Figure 7-28(a) for the reader's convenience. The resulting output chart is shown in Figure 7-28(b). The construction of the chart is performed by inspecting the outputs of each stable state in the RFC. For example, the outputs of state 1 is (00); therefore, (00) is entered in the location corresponding to state 1. Examining the RSC further, it can be noted that unstable state 2 can result from both states 1 and 5. Examining states 1 and 2, it can be seen that output state R is the same for both states, and therefore, a 0 is assigned to this particular output. On the other hand, due to the difference between states 1 and 2, output Q is assigned a "don't care." However, since output Q remains the same during the transition from state 5 to state 2, output Q is assigned a 1. Similarly, an examination of the RSC shows that state 4 may occur following states 2 and 9. During these transitions, both outputs anticipate changes and, therefore, "don't cares" are assigned for the particular map location. The completed Output Chart can be observed in Figure 7-28(b).

Y_1	Y_2	I_1	I_2	I_3	I_4
0	0	(1)	2	(5)	(3)
0	1	(6)	(2)	4	(9)
1	1	(11)	(8)	(4)	7
1	0	1	(12)	(10)	(7)

(a)

Y_1	Y_2	I_1	I_2	I_3	I_4
0	0	00	01	01	00
0	1	10	01	--	01
1	1	11	00	10	11
1	0	--	11	11	11

(b)

Figure 7-28 The construction of the Output chart for the RSC in Figure 7-10: (a) The Operational Flow Chart; (b) the Output Chart.

SEQUENTIAL LOGIC DESIGN—CLASSICAL SYNTHESIS

$Y_1 Y_2$ \ AB	00	01	11	10
00	00	01	01	00
01	10	01	--	01
11	11	00	10	11
10	--	11	11	11

(R, Q)

Figure 7-29 The Karnaugh map of the outputs (R, Q).

The Karnaugh output map is developed in this case by assuming that the input states are (00), (01), (11), (10), respectively for I_1, I_2, I_3, and I_4. The Karnaugh map for the example problem is shown in Figure 7-29. The network output equations are therefore:

$$R = AY_1 + Y_1 Y_2' + A'B'Y_2 \qquad (7\text{-}13)$$
$$Q = BY_1' + AY_1' Y_2 + B'Y_1 + Y_1 Y_2' \qquad (7\text{-}14)$$

Note that the output equations as well as for the memory equations can be derived and implemented using all the rules and algorithms presented in earlier chapters.

7-9 SEQUENTIAL NETWORK HAZARDS

Basically, all hazards that can prevail in combinational circuitry can occur in sequential networks. This is due to the fact that, actually, a sequential network consists of combinational circuits integrated with the added memory devices. The problem can become much more severe, however, since transient changes in some secondary outputs may cause permanent incorrect changes of the output. Such situations are referred to as "steady state hazards," as the outputs assume false outputs "permanently." The identification and elimination of these types of hazards, which can also occur due to other causes, will be the main objective of this section.

In general, hazards in sequential logic systems can be classified into three major classes:

1 Hazards due to prevailing combinational logic hazards.
2 Hazards due to erroneous synthesis.
3 Hazards due to races between input and secondary variables.

Combinational logic synthesis hazards have been treated in Chapter 6 and, therefore, no attempt will be made here to duplicate the discussion of this type of hazard. Under the second category, there are various logical complications that are actually caused by erroneous synthesis, i.e., by actions that are actually not permissible.

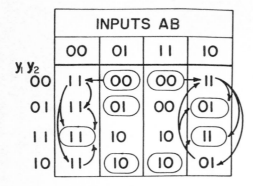

Figure 7-30 Excitation map showing races.

Although these complications should not be classified as hazards, they are often referred to as hazards and, therefore, will be discussed in this section.

The first type of "hazard" due to erroneous synthesis to be discussed is the "race" condition. A race occurs when it is necessary to switch more than one memory at the same time. Since a memory element, like all logic elements, requires some finite amount of time to complete the transition, the problem becomes a "who will win the race" type problem. If a race always terminates in this desired state, then the race is termed a *noncritical race*; on the other hand, if there is a possibility that a race will terminate in an undesired location, the race is called a *critical race*. In Figure 7-30, the Excitation map shows several races. Observe that in the first column, the $Y_1 Y_2 = (00)$ location, both memories are unstable and require a transition to the third row. If, for example, Y_1 "wins" the race, the system finds itself in the fourth row, which requires an additional transition to the third row due to the unstable condition of Y_2. Similarly, if Y_2 is switched earlier than Y_1, the third row is reached because the unstable condition of row 2 requires further transition to row 3. Therefore, the race condition represented in the first column of Figure 7-30 is a noncritical race.

A critical race is illustrated in the fourth column. For example, if the system is initially at the first row, the unstable nature of Y_1 and Y_2 would indicate the intention of the system to switch to the third row. However, if in some cases Y_2 occurs before Y_1, then the system will find itself stable at the second row. On the other hand, if Y_1 wins the race, the system will again be involved in another race condition at the fourth row, which requires a transition to the second row. If the system is in the fourth row and Y_2 is finally actuated, then the system will luckily be in its correct, intended state in row 3. It should be obvious that critical races are not desired in a system, as they cause erroneous operations. However, noncritical races may prove to be an important facet of network simplification. For example, for the problem depicted in Figure 7-9, the OFC can be constructed as shown in Figure 7-31. If the use of noncritical races is ignored, the corresponding Excitation map can be constructed as shown in Figure 7-32. Utilizing standard SET and RESET memory elements, the SET and RESET equations are given as follows:

$$\text{SET}_{Y_1} = ABY_2 + A'BY_3 \qquad (7\text{-}15)$$

SEQUENTIAL LOGIC DESIGN—CLASSICAL SYNTHESIS

	Inputs (AB)			
$Y_1 Y_2 Y_3$	00	01	11	10
0 0 0	(1)	2	(7)	5
0 0 1	1	2	(3)	5
0 1 1	—	—	—	—
0 1 0	1	2	6	(5)
1 1 0	1	2	(6)	4
1 1 1	—	—	—	4
1 0 1	1	(2)	3	(4)
1 0 0	—	—	—	—

Figure 7-31 The Operational Flow Chart.

$$\text{RESET}_{Y_1} = A'Y_2 + ABY_3 + A'B' \tag{7-16}$$
$$\text{SET}_{Y_2} = AB'Y_3' \tag{7-17}$$
$$\text{RESET}_{Y_2} = A'Y_1' + Y_3 \tag{7-18}$$
$$\text{SET}_{Y_3} = A'BY_2' + AB'Y_1 \tag{7-19}$$
$$\text{RESET}_{Y_3} = B'Y_1' \tag{7-20}$$

Utilizing all types of two-input elements, the simplest network that can be constructed using the above equations results in the sixteen-element network shown in Figure 7-33.

On the other hand, if the simplification features offered by the insertion of noncritical races are considered, the resulting Excitation map is as shown in Figure 7-34. To illustrate the insertion of noncritical races, consider the transitions indicated in column 1 at all unstable locations. In order for these unstable states to transition optimally to their corresponding stable state, the memory excitation associated with the stable state must be assigned to each unstable state. This, of course, will cause a race condition throughout the whole first column of the map. Therefore, all rows should be inspected to determine whether it is possible for the system to end up accidentally in an undesired row. When the system can reach "–" locations of the chart, then such "–" should be assigned proper excitations in order

	Inputs (AB)			
$Y_1 Y_2 Y_3$	00	01	11	10
0 0 0	000	001	000	010
0 0 1	000	101	001	000
0 1 1	—	—	—	—
0 1 0	000	000	110	010
1 1 0	010	010	110	111
1 1 1	—	—	—	101
1 0 1	001	101	001	101
1 0 0	—	—	—	—

Figure 7-32 The Excitation map, without races.

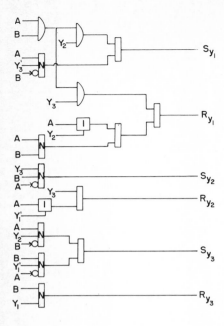

Figure 7-33 SET-RESET circuit without inserted races.

to achieve the desired transition. Obviously, when the insertion of races can cause an unwanted situation, the particular race condition must not be used. In the example under discussion, the eighth row can accidentally be reached by the system, and, therefore, the excitation (000) is placed in this particular location. Similarly, races are inserted in the second and fourth columns, as can be observed in Figure 7-34. The associated SET and RESET equations resulting from the insertion of the noncritical races are as follows:

$$\text{SET}_{Y_1} = A'B + BY_2 \tag{7-21}$$
$$\text{RESET}_{Y_1} = A'B' + ABY_3 \tag{7-22}$$
$$\text{SET}_{Y_2} = AB'Y_1' \tag{7-23}$$

	Inputs (AB)			
$Y_1 Y_2 Y_3$	00	01	11	10
0 0 0	000	101	000	010
0 0 1	000	101	001	010
0 1 1	—	101	—	010
0 1 0	000	101	110	010
1 1 0	000	101	110	101
1 1 1	—	101	—	101
1 0 1	000	101	001	101
1 0 0	000	101	—	101

Figure 7-34 The Excitation map, containing races.

SEQUENTIAL LOGIC DESIGN—CLASSICAL SYNTHESIS

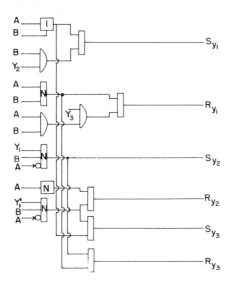

Figure 7-35 SET-RESET circuit with races.

$$\text{RESET}_{Y_2} = A' + AB'Y_1 \tag{7-24}$$
$$\text{SET}_{Y_3} = A'B + AB'Y_1 \tag{7-25}$$
$$\text{RESET}_{Y_3} = AB'Y_1' + A'B' \tag{7-26}$$

These expressions can be implemented as shown in the thirteen element circuit in Figure 7-35.

Other types of complications caused by improper synthesis are of minimum importance. These include "false output hazards," "cycles," and "lockups." A false output hazard occurs when an output state is the same for two stable states in a transition, but has a different output assigned for the unstable or transitory state. Therefore, such "hazard" will not occur when the rules for assigning outputs (Section 7-6) are used. Cycles are situations where no stable state exists in a column, while lockups are characterized by a group of stable states which locks out the other states from the operation of the system. Both of these phenomena are

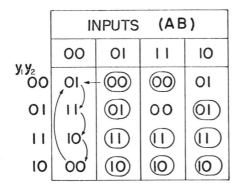

Figure 7-36 Excitation map showing a cycle.

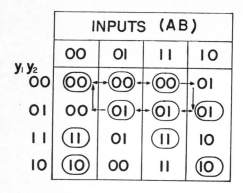

Figure 7-37 Excitation map showing a lockup.

illustrated in Figures 7-36 and 7-37. Note that cycles and lockups may have been inserted deliberately in order to satisfy some particular need.

Under the third category are classified the critical types of hazards which are characteristic only of sequential networks. These hazards are characterized by races between inputs and memory signals. A hazard of this type that is encountered most often is the "essential hazard." The existence of these hazards were initially recognized by Unger who defined this type of hazard as follows:

> A sequential function contains an essential hazard if there exists a stable state, S_o, and an input variable A such that, starting with the system at S_o, three consecutive changes in A bring the system to a state other than the one arrived at after the first change in A.

The circuit represented by the OFC in Figure 7-38 shows the existence of the essential hazard. In this table, it can be observed that one change of A ($0 \rightarrow 1$) will cause the circuit to go to state 2, while three consecutive changes will bring the circuit to state 4. It should be noted that the flow table in Figure 7-38 contains four essential hazards, namely the ones starting at states 1, 2, 3 and 4. In order to illustrate the influence of this type of hazard, consider the secondary circuit shown in Figure 7-39. Assume that the system is currently at state 1, while A changes from $0 \rightarrow 1$. At this instant, Y_1 and Y_2 are both in the "reset" condition. Following the activation of A, the AND-valve state 3 energizes, resulting in the "setting" of the Y_2 memory element. If, because of the delay associated with the

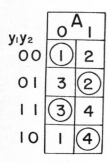

Figure 7-38 Flow chart containing essential hazards.

SEQUENTIAL LOGIC DESIGN—CLASSICAL SYNTHESIS

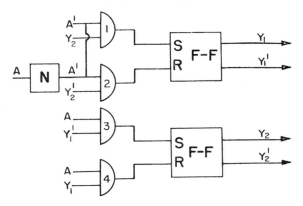

Figure 7-39 The secondary circuit realization.

"NOT-ing" of A, A' does not drain fast enough, then AND-valve (1) might be energized, resulting in the switching of memory element Y_1. At this step, Y_2 detects an unstable state 4, which results in the resetting of Y_2 by means of AND-valve number (4). Assuming that the signal A' has been drained completely at this instant, the system will remain stable at state 4, which is not the intended state. This input/memory race can be further illustrated by the signal scheduling shown in Figure 7-40.

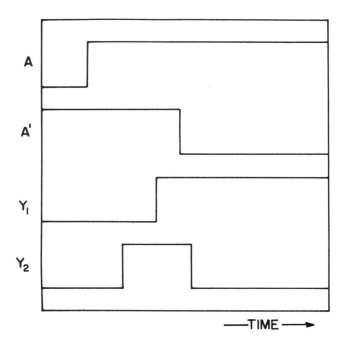

Figure 7-40 Signal scheduling, exhibiting an essential hazard.

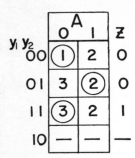

Figure 7-41 "d-trio" causing a transient hazard.

An almost similar situation occurs when a "d-trio" exists in the system, as illustrated in Figures 7-41 and 7-42. Unger defined the "d-trio" as follows:

> Three rows of a flow table, r_1, r_2, r_3, constitute a d-trio if, for some pair of input states L and R, differing only in the value of one input variable, the next-state entries in L are r_1, r_3, and r_3 in Rows r_1, r_2, and r_3, respectively, and the next-state entries in R are r_2 for all three rows.

The hazard conditions resulting from a d-trio can be understood by referring to Figure 7-43 and its associated LSC shown in Figure 7-41. At first glance, it may seem that the existence of a d-trio will not constitute an erroneous operation. However, in particular situations, different types of hazards may exist due to the presence of d-trios in the networks. If the output at state 3 is 1, while at the other states it is 0 or OFF, then if the network is at state 1, a change in A will bring the network eventually to state 2. However, if the race between the input and Y_2 is won by Y_2 as illustrated in Figure 7-40(b), then the network transition might be

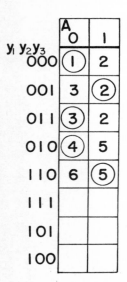

Figure 7-42 "d-trio" causing a third-order hazard.

SEQUENTIAL LOGIC DESIGN—CLASSICIAL SYNTHESIS

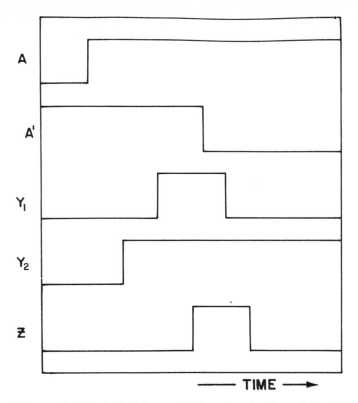

Figure 7-43 Signal scheduling, exhibiting a transient hazard due to "d-trio."

$1 \xrightarrow{\cdot} 3 \rightarrow 2$, which obviously will result in a transitional output (transient hazard) when the network is temporarily at State 3.

A different situation exists in the flow table shown in Figure 7-42. For example, assume the initial state to be state 1 and the input changes from $0 \rightarrow 1$. At this moment, the circuit Y_3 senses the transition and switches Y_3 to 1. Consider, for instance, the case where the SET-circuit of Y_2 senses A' and Y_3 (A has not yet been drained completely), and sets Y_2 to 1; then the network assumes state 3 as its "intended" state. Furthermore, it is possible that at this instant of time the input A has not reached Y_2, while Y_1 may have sensed Y_2 and A but not Y_3. In this case, Y_1 will assume that the system is actually in unstable state 5, which results in the activation of Y_1 and sends the system to stable state 5. No particular name has been assigned to this hazard; however, as three memory variables are involved in the transition, this hazard is classified as a "third-order hazard." Consequently, the essential hazard discussed earlier can be classified as a "second-order hazard." As there still exist many situations that constitute second, third...Nth order hazards, no attempt will be made to further illustrate these types of hazards.

The reader may note that the hazards categorized under the third classification

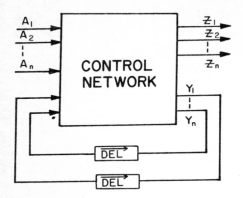

Figure 7-44 General scheme of hazard elimination using delay elements.

prevail when the "races" between the changes of the input and the secondary variables are "won" by the secondaries. Consequently, a remedy for this particular type of hazard can be achieved by simply delaying the feedback signal of the memory elements. The amount of the delay can be adjusted to cause the output of the memory to arrive at a later time after the transition of the changing input variable has been completed. It should be noted, however, that adjusting this delay factor should be made such that the delay is minimal, due to the fact that it influences the speed of response of the network. A generalized scheme for a hazard elimination using delay elements is shown in Figure 7-44.

7-10 COMPUTER-AIDED SYNTHESIS

Synthesizing small problems manually can be accomplished in an expedient manner. However, large problems are time-consuming and the probability for error becomes critical. Hence, computers are desirable to provide convenience and avoid errors

Table 7-6 Example Input Deck to CLASYN

```
        1         2         3         4
1234567890123456789012345678901234567890123456789

 10   8   3   1   1   0
  1   0   0  -1   0   0   2   9   0   0   7
  0   1   0   3   0   0  -2   0   0   0   0
  0   0   0  -3   4   0   0   0   0   0   0
  1   0   0   0  -4   5   0   0  10   8   0
  0   1   0   0   6  -5   0   0   0   0   0
  0   0   1   1  -6   0   0   0   0   0   0
  0   1   0   1   0   0   0   0   0  -7
  1   0   1   0   4   0   0   0  -8   7
  0   1   0   0   0   0   2  -9   0   0
  0   1   0   0   0   5   0   0 -10   0   0
```

SEQUENTIAL LOGIC DESIGN—CLASSICAL SYNTHESIS

Table 7-7 Resulting Circuit Equations for Example Problem of Table 7-6

Circuit Equations:

	Y_1	Y_2	Y_3	Y_4	Y_5	A	B	C	D	
$Y_1 =$										
	—	—	—	—	—	1	—	1	—	
	—	—	—	—	—	0	1	0	—	
	1	—	—	—	—	—	0	0	—	
	1	—	—	—	—	1	0	—	—	*
	1	—	—	—	—	0	—	0	—	*
$Z_1 =$										
	1	—	—	—	—	1	0	—	—	
	0	—	—	—	—	0	0	0	—	
$Z_2 =$										
	—	—	—	—	—	—	1	—	—	
	0	—	—	—	—	—	—	1	—	
$Z_3 =$										
	0	—	—	—	—	1	0	—	—	
	—	—	—	—	—	1	0	1	—	

*Hazard-elimination term.

within the synthesis. A program referred to as CLASYN has been developed based on the Classical Synthesis Method.

The program is designed to accept a logic specification in LSC form, where the outputs are related only to the stable states of the system. The program then performs all the necessary manipulations and produces the circuit equations which can be implemented with the desired logic hardware. The input to the program is illustrated in Table 7-6. The first card describes the size of the problem—the number of rows, the number of the columns in the next state subtable, the number of outputs, and the type of output that is requested. In the following cards, the LSC is entered, row by row, entering negative numbers for stable state conditions. For example, (−10) represents the stable state 10, while the positive number (10) refers to the unstable condition of the machine state 10. The output of the program can be observed in Table 7-7.

PROBLEMS

1. From the LSC of Figure P7-1, obtain the minimal-state machine by eliminating the redundant states.
2. Form the RSC of the LSC shown in Figure P7-2.
3. Derive the RSC of the LSC shown in Figure P7-3.

Outputs		Input states			
P	Q	I_1	I_2	I_3	I_4
0	0	(1)	2	5	—
0	1	1	(2)	—	8
1	0	7	—	(3)	9
1	1	(4)	6	3	—
1	0	1	—	(5)	9
0	1	4	(6)	—	8
0	0	(7)	2	3	—
1	1	—	6	3	(8)
1	1	—	2	5	(9)

Figure P7-1

Outputs		Input states			
P	Q	I_1	I_2	I_3	I_4
0	0	(1)	2	6	3
0	1	5	(2)	10	8
—	1	1	—	10	(3)
0	1	1	(4)	9	3
0	0	(5)	4	10	7
1	1	—	4	(6)	3
0	1	5	—	6	(7)
1	1	1	—	9	(8)
1	0	—	2	(9)	3
1	—	—	2	(10)	7

Figure P7-2

Outputs		Input states			
P	Q	I_1	I_2	I_3	I_4
0	1	(1)	2	—	3
0	—	4	(2)	11	—
—	0	6	—	12	(3)
0	1	(4)	5	—	7
0	1	1	(5)	12	—
1	0	(6)	2	—	8
1	0	6	—	11	(7)
0	0	9	—	13	(8)
1	0	(9)	10	—	3
0	0	4	(10)	12	—
1	1	—	5	(11)	8
1	—	—	2	(12)	3
1	0	—	10	(13)	7

Figure P7-3

SEQUENTIAL LOGIC DESIGN—CLASSICAL SYNTHESIS

4 Obtain the Minimal Row Specification Charts (MSC) of the RSC's obtained in Problems **1**, **2**, and **3**.

5 From the RSC shown in Figure P7-5, form the MSC.

Outputs		Input states			
P	Q	I_1	I_2	I_3	I_4
0	1	(1)	2	—	4
0	0	1	(2)	3	—
0	0	—	2	(3)	4
1	1	7	—	5	(4)
0	1	—	6	(5)	—
1	0	1	(6)	—	—
0	1	(7)	8	—	10
1	1	7	(8)	9	—
0	1	—	2	(9)	10
1	0	7	—	11	(10)
0	0	—	2	(11)	10

Figure P7-5

6 Formulate the MSC from the RSC in Figure P7-6.

Outputs			Input states							
P	Q	R	I_1	I_2	I_3	I_4	I_5	I_6	I_7	I_8
1	0	0	(1)	4	—	2	—	—	—	7
0	1	0	3	—	5	(2)	9	—	—	—
0	0	0	(3)	6	—	—	—	—	—	—
1	0	0	3	(4)	5	—	—	—	8	—
1	1	1	—	6	(5)	—	—	10	—	—
1	1	0	1	(6)	—	—	—	—	—	—
0	0	0	3	—	—	—	—	—	—	(7)
1	0	1	—	6	—	—	—	10	(8)	7
1	0	0	—	—	—	2	(9)	10	—	7
0	1	0	—	—	5	—	9	(10)	8	—

Figure P7-6

7 Obtain the Operational Flow Chart (OFC) of the MSC's obtained in Problems **4**, **5**, and **6**.

8 From the MSC shown in Figure P7-8, obtain the Operational Flow Chart.

9 Develop the minimal row OFC from the MSC shown in Figure P7-9.

10 For the OFC's obtained in Problems **8** and **9**, perform the following:
   a Formulate the standard SET/RESET expressions.
   b Obtain minimal SET/RESET expressions for implementation using the FLUID MEMORY element represented in Table 7-4.
   c Obtain minimal SET/RESET expressions for implementation using the detented Flip-Flop. Show the degree of simplification achieved by the three approaches using two-input, logic elements for implementation.

Inputs (a, b)			
00	01	11	10
(1)	2	(5)	6
(7)	2	5	6
7	2	10	(6)
1	2	(10)	11
(3)	9	(8)	4
3	(2)	8	(11)
3	(9)	5	(4)

Figure P7-8

Inputs (a, b)			
00	01	11	10
(1)	2	(3)	4
(5)	(2)	11	12
—	—	3	(8)
14	13	(11)	(4)
5	(6)	(7)	8
1	6	(9)	(10)
14	—	—	(12)
(14)	(13)	7	10

Figure P7-9

11 From the MSC's developed in Problems 4 and 5 and their related RSC's, obtain the network equations (SET, RESET, and output equations) by using standard memory devices. Assume that the input states I_1, I_2, I_3, and I_4 are 00, 01, 11, 10, respectively.

12 From the MSC developed in Problems 6 and its related RSC, obtain the network equations by assuming its input states are 000, 001, 011, 010, 110, 111, 101, 100 for I_1, I_2, ... I_8, respectively. The network is to utilize FLUID MEMORY elements as shown in Table 7-4.

13 The LSC of Figure P7-13 should be synthesized and implemented using detented FLIP-FLOP elements as their memory devices. Obtain the minimal representation of the network.

Outputs		Inputs (a, b)			
P	Q	00	01	11	10
1	0	(1)	3	—	2
1	0	4	—	5	(2)
1	0	1	(3)	—	—
1	0	(4)	3	—	2
0	1	—	—	(5)	6
0	1	1	—	—	(6)

Figure P7-13

S	R	Y*	Y
0	0	0	0
0	0	1	1
0	1	1	0
0	1	0	0
1	1	0	0
1	1	1	0
1	0	1	1
1	0	0	1

Figure P7-14

14 Assume memory elements have characteristics shown by the truth table given in Figure P7-14. Propose an algorithm for deriving the minimal SET and RESET circuits. Apply the element to the problems discussed in Problem 10 and compare results.

15 Identify all essential hazards and d-trios in the MSC of Problem 8.

16 Identify all essential hazards and d-trios in the OFC resulting in Problems 7 and 9.

DEFINITIONS

Adjacency A relationship existing between two stable states such that there is only one describing variable which is different in value (0, 1).

Memory excitation A stimulation of the secondary or memory system.

Memory or secondary state The collective state or condition of the memory elements.

Memory transition The stepwise change of the memory state.

Minimal Row Specification Chart A specification chart that describes the logic operation of a machine with the minimal number of rows.

Operational Flow Chart A specification chart that represents the operational logic of a machine and which satisfies the adjacency requirements of its memory states.

Reduced Specification Chart A logic specification chart containing the minimal number of machine states.

Row compatibility Two rows are compatible if they can be merged into one row.

State equivalency Two machine states are equivalent if they can be represented by one machine state.

REFERENCES

Armstrong, D. B., Friedman, A. D., and Menon, P. R., "Realization of Asynchronous Sequential Circuits without Inserted Delay Elements," *IEEE Trans. on Electronic Computers*, Vol. EC-17, pp. 129–134, Feb. 1968.

———, "Design of Asynchronous Circuits Assuming Unbounded Gate Delays," *IEEE Trans. on Electronic Computers*, Vol. EC-18, No. 12, pp. 1110–1120, Dec. 1969.

Aufenkamp, D. D., "Analysis of Sequential Machines II," *Trans. IRE Professional Group on Electronic Computers*, Vol. 7–8, pp. 299–306, Dec. 1958.

———, and Hohn, F. E., "Analysis of Sequential Machines," *IRE Trans. on Electronic Computers*, pp. 276–285, Dec. 1957.

Beizer, B. S., and Leibholz, S. W., "Engineering Applications of Boolean Algebra: Part 5: "Designing Sequaential Circuits," *Electrical Manufacturing*, Vol. 62, pp. 67–69, 298, 300, Sept. 1958.

Biswas, N. N., "State Minimization of Incompletely Specified Sequential Machines," *IEEE Trans. on Electronic Computers*, Vol. EC-23, No. 1, pp. 80–84, Jan. 1974.

Dietmeyer, D. L., *Logic Design of Digital Systems*. Boston: Allyn and Bacon Inc., 1970.

DeMoss, D. M., "Criteria for the Design of Fast, Safe, Asynchronous Sequential Fluidic Circuits," Ph.D. Dissertation, Oklahoma State University, Stillwater, Oklahoma, 1967.

Eichelberger, E. B., "Obtaining a Minimal State Compressed Flow Table," Digital Systems Laboratory, Princeton University Technical Report No. 20, April, 1962.

———, "Hazard Detection in Combinational and Sequential Switching Circuits," *Proc. Fifth Annual Symp. Switching Circuit Theory and Logical Design*, pp. 111–120, Nov. 1964.

Fitch, E. C., "Fluid Logic," Teaching Manual Published by The School of Mechanical and Aerospace Engineering, Oklahoma State University, Stillwater, Oklahoma, pp. 175-189, 1966.

———, "Synthesis of Fluid Logic Systems," ASME Paper No. 70-DE-47, Presented at the Design Engineering Conference and Show, Chicago, May 1970.

Friedman, A. D., "Feedback in Asynchronous Sequential Circuits," *IEEE Trans. on Electronic Computers*, Vol. EC-15, No. 5, pp. 740-749, Oct. 1966.

———, and Menon, P. R., "Synthesis of Asynchronous Sequential Circuits with Multiple Input Changes," *IEEE Trans. on Electronic Computers*, Vol. EC-17, No. 6, pp. 559-566, June 1968.

Ginsburg, S., "On the Minimization of Superfluous States in a Sequential Machine," *Journal of the Assoc. of Computing Machinery*, Vol. 6, pp. 259-282, April 1959.

———, "A Technique for the Reduction of a Given Machine to a Minimal State Machine," *IRE Trans. on Electronic Computers*, Vol. 8, pp. 346-355, September 1959.

———, "Synthesis of Minimal State Machines," *IRE Trans. on Electronic Computers*, Vol. EC-8, pp. 441-449, Dec. 1959.

Givone, D. D., *Introduction to Switching Circuit Theory*. New York: McGraw-Hill Book Company, 1970.

Goedecke, Wolf-Dieter, "Rechnergestüzte Synthese asynchroner sequentieller Schaltkreise in der Fluidik," Doktor-Ingenieurs Dissertation, Rheinisch Westfalischen Technischen Hochschule Aachen, June 1974.

Grasselli, A., and Luccio, F. "A Method for Minimizing the Number of Internal States in Incompletely Specified Sequential Networks," *IEEE Trans. on Electronic Computers*, Vol. EC-14, pp. 350-359, June 1965.

Hackbart, R. R., and Dietmeyer, D. L., "The Avoidance and Elimination of Function Hazards in Asynchronous Sequential Circuits," *IEEE Trans. on Electronic Computers*, Vol. EC-20, No. 2, pp. 184-189, Feb. 1971.

Hlâvicka, J., "Essential Hazard Correction Without the Use of Delay Elements," *IEEE Trans. on Electronic Computers*, Vol. EC-19, No. 3, pp. 232-238, March 1970.

Huffmann, D. A., "Synthesis of Sequential Switching Circuits," *Journal of the Franklin Institute*, Vol. 257, Nos. 3 and 4, pp. 161-190, pp. 275-303, March-April 1954.

———, "The Design and Use of Hazard-Free Switching Networks," *Journal of Associate Computing Machinery*, Vol. 4, No. 1, pp. 47-62, Jan. 1957.

Kohavi, A., *Switching and Finite Automata Theory*. New York: McGraw-Hill Book Co., 1970.

Lerner, S. B., "Hazard Correction in Asynchronous Sequential Circuits," *IEEE Trans. on Electronic Computers*, Vol. EC-14, pp. 265-267, April 1965.

Luccio, F., "Extending the Definition of Prime Compatibility Classes of States in Incomplete Sequential Machine Reduction," *IEEE Trans. on Electronic Computers*, Vol. EC-18, No. 6, pp. 537-540, June 1969.

Marcus, M. P., *Switching Circuits for Engineers*. Englewood Cliffs, N.J.: Prentice-Hall, Inc., 1962.

———, "Relay Essential Hazards," *IEEE Trans. on Electronic Computers*, Vol. EC-12, pp. 405-407, Aug. 1963.

Paull, M. C., and Unger, S. H., "Minimizing the Number of States in Incompletely Specified Sequential Functions," *IRE Trans. on Electronic Computers*, Vol. 8, pp. 356–366, Sept. 1959.

Quine, W. V., "The Problem of Simplifying Truth Functions," *American Math. Monthly*, Vol. 59, pp. 521–531, Oct. 1952.

Reed, I. S., "On the State Reduction of Asynchronous Circuit on the Paull-Unger Method," *IEEE Trans. on Electronic Computers*, Vol. EC-14, pp. 262–265, April 1965.

Saucier, G., "Next State Equations of Asynchronous Sequential Machines," *IEEE Trans. on Electronic Computers*, Vol. EC-21, No. 4, pp. 397–399, April 1972.

Surjaatmadja, J. B., "A Computer-Oriented Method for Boolean Simplification and Potential Hazard Elimination," Fluid Power Research Conference, Rept. No. R73-FL-2, Fluid Power Research Center, Oklahoma State Univ., Stillwater, Oklahoma, 1973.

_____, "A Generalized Method for Synthesizing Optimal Fluid Logic Networks," Fluid Power Research Conference, Rept. No. R73-FL-3, Fluid Power Research Center, Oklahoma State University, Stillwater, Oklahoma, 1973.

_____, "Essential Hazard Elimination: The Deprivation of its Essentiality in Sequential Circuits," Eighth Annual Fluid Power Research Conference, Oklahoma State University, Stillwater, Oklahoma, October, 1974.

_____, "A Graphical Approach for the Development of a Near-Minimal Row Operational Flow Table," Fluid Power Research Conference, Paper No. P75-63, Fluid Power Research Center, Oklahoma State University, Stillwater, Oklahoma, October, 1975.

Tracey, J. H., "Internal State Assignments for Asynchronous Sequential Machines," *IEEE Trans. on Electronic Computers*, Vol. EC-15, No. 4, pp. 551–560, Aug. 1966.

Unger, S. H., "Hazards and Delays in Asynchronous Sequential Switching Circuits," *IRE Trans. on Circuit Theory*, Vol. CT-6, pp. 12–56, March 1959.

_____, "Asynchronous Sequential Switching Circuits With Unrestricted Input Changes," *IEEE Trans. on Electronic Computers*, Vol. EC-20, No. 12, pp. 1437–1444, Dec. 1971.

Vadasz, A. F., and Fitch, E. C., "A Nodal Diagram Technique for Achieving Minimal Memory Logic Systems," Fluid Power Research Conference, Paper No. P75-58, Fluid Power Research Center, Oklahoma State University, Stillwater, Oklahoma, October, 1975.

Wood, P. E., *Switching Theory*. New York: Lincoln Lab Publication, McGraw-Hill Book Co., pp. 279–330, 1968.

Yang, C. C., "Closure Partition Method for Minimizing Incomplete Sequential Machines," *IEEE Trans. on Electronic Computers*, Vol. EC-22, No. 12, pp. 1109–1122, Dec. 1973.

Chapter 8

Sequential Logic Design—Nonclassical

8-1 QUEST FOR OPTIMAL SYNTHESIS

Since 1954, when Huffman introduced the first technique for synthesizing sequential networks, there have been repeated attempts to discover a new method which would have a higher degree of optimality—in synthesis time, network response, and/or circuit complexity. This quest has not been entirely fruitless, because techniques have been created which have proven superior for a specific class of systems; e.g., a deterministic type. Without a doubt, the search will continue as long as new applications demand the advancement of fluid logic technology or the acme of synthesis is attained.

Specifically, the motivation for pursuing alternative methods stems from the need to:

1 Achieve a more minimal network (one containing fewer logic elements).
2 Reduce the expertise needed to perform the synthesis.
3 Create a method possessing a simpler format for a given type of network.
4 Formalize a user-oriented computer algorithm.

The methods reported in the literature invariably reflect objectives which were formalized due to such motivation. The methods described in this chapter represent

SEQUENTIAL LOGIC DESIGN—NONCLASSICAL

the best attempts in achieving a viable new method. Although some are presented which fall short of their intended goal, they all contribute important facets which can help future investigators attain possible new objectives.

The methods presented here are identified by both the most distinguishing feature of the method and its author; viz.,

- Change Signal Method—J. H. Cole
- Total Signal Method—G. E. Maroney
- State Matrix Method—R. L. Woods
- Transition Table Method—R. M. H. Cheng and K. Foster
- State Diagram Method—P. I. Chen and Y. H. Lee

After studying each method, the reader will recognize that each technique has its own distinct advantages over the other and that the degree of optimality achieved depends upon the types of problems that are considered.

8-2 THE CHANGE SIGNAL METHOD

In 1968, J. H. Cole realized that in a deterministic logic system, the input signal which changed last is the most influential factor in the activation of an output at a particular instant. He also realized that the Operations Table, a table that at that time had found usefulness only in logic analysis, also could serve as a powerful synthesis tool in the area of fluid logic. Utilizing these concepts, he developed his Change Signal Method, which reflects the following advantages relative to the Classical Synthesis Method:

1. The procedure is quite simple to use. All synthesis operations can be performed in a single table.
2. The procedure is capable of solving relatively large, finite-state machines involving a large number of input variables.
3. The procedure produces a network with prepared signal paths, thus yielding the fastest logic circuit possible.

The reader should realize that the method is tailored for synthesizing only deterministic, feedback-type sequential networks having a particular type of input-output circuit, as illustrated in Figure 8-1. Minor changes in the procedure are required when other input-output circuit configurations are used. Absolutely no irregular conditions are tolerated by the change signal approach.

The Change Signal Method incorporates a tabular representation which is known as the Operations Table (Cole also referred to it as merely the "synthesis table"). All the necessary synthesis steps are performed and recorded on the table itself; hence, the completed Operations Table contains the concise operational specification of the circuit as well as the logic structure of the completed circuit itself. The construction of the preliminary form of the operations table was presented in Chapter 5.

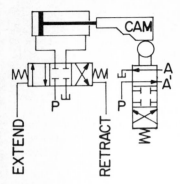

Figure 8-1 Typical input-output circuit.

The procedure for the method can be conveniently outlined as follows:

1. Construct the preliminary form of the operations table as required by the logic specification of the problem.
2. Inspect the table for the existence of conflicting input-output relationships. A conflicting condition is represented by a situation where the system should have different outputs following a similar input change at different periods of time.
3. Augment the secondary states by the use of memory elements (which are indicated by Y's) in order to eliminate any conflicting conditions. Memory augmentation is performed in a Gray Code fashion. If the number of memory states that can be achieved by the augmented memories exceeds the number of nonunique, conflicting input states, assign more than one memory state to the last nonunique conflicting states, thus forming memory states having fewer memory variables.
4. Augment auxiliary Y-memory valves to establish uniqueness of each machine state.
5. Select the SET and RESET signals for the mandatory Y-memory elements. Reset each valve in the row immediately preceding the row in which the valve must be OFF. However, if the immediately preceding row is represented by an auxiliary Y-valve, try to reset the valve in all locations where this particular auxiliary Y-valve is represented. If conflicting conditions occur, try the next earlier row location to see if the use of this auxiliary Y-valve can be avoided. Note that this row selection can only be made amidst the group of rows immediately preceding the RESET row, where the associated input signal is OFF. If the use of an auxiliary Y-valve cannot be avoided, classify this valve as mandatory. The signal selection for the SET condition is performed in the same manner. However, as the RESET signal has been selected at this moment, any conflicts between the SET and RESET signals should be avoided. In the selection of the SET and RESET signals, it is recommended that the above trials are made after all "unavoidable" or "favorable" SET/RESET signals have been determined.
6. Utilize "shut-off" valves (represented by W's) for terminating unwanted or persistent output signals.
7. Select the necessary SET and RESET signals for the W-valves. The resetting of the W-valve should be made in the row immediately preceding the row in which the W-valve must be OFF, whenever possible. If conflicting outputs occur in

SEQUENTIAL LOGIC DESIGN—NONCLASSICAL

adjacent rows, reset the W-valve in the row where the W-valve must be OFF, whenever possible. The use of auxiliary memory elements should also be avoided. If the use of such auxiliary memory valves cannot be avoided, classify them as mandatory and determine their SET and RESET signals. The SET signal for the W-valves is preferably performed in the row immediately preceding the row where the particular W-valve must be ON. However, if this causes a contradiction with the RESET signal, or causes the utilization of an auxiliary Y-memory valve, try the next earlier row and so forth until a satisfactory SET signal is reached.

8 Obtain the circuit equations. Each output signal consists of the appropriate input signal, the memory signal, and the shut-off W signals of each of the related rows. Note that only the uncomplemented W's are considered. The SET and RESET signals consist only of the input signals and the memory signals.

9 If a prepared path network is desired, replace each OR-element with a memory valve. (Cole assumed that all logic functions are achieved by using only three-way and four-way valves. When other types of valves are utilized, this step may be inappropriate.) The SET and RESET locations of each of these "new" OR elements are selected immediately preceding the occurrence of the particular augmented output signal. It should be realized, however, that this step is not always possible when some nonconventional input-output circuits (other than those presented in the circuit shown in Figure 8-1) are incorporated.

In order to illustrate the procedure, consider the problem involving two cylinders which are required to perform the following sequence: $A, B, \bar{B}, \bar{A}, B, A, \bar{A}, \bar{B}$, where A and $\bar{A}$ are used to represent the extension and retraction of cylinder A, respectively. The input and output circuits for both cylinders A and B are shown in Figure 8-1. Input variables a and b are used to indicate the retraction and the extension of cylinders A and B, respectively. The preliminary Operations Table is illustrated in Table 8-1. In the first column, the state numbers are listed, followed by the sequence of events in the second column, and the changed signals in the third column (note that the total signals are not listed in this table). In these types of problems, where the inputs merely indicate the completion of each circuit operation, the changed inputs are represented by the sequence of events but delayed by one state. Thus, as the first event is A, the changed state at the second row is a, and so forth. Furthermore, to simplify the synthesis, in the next column the

Table 8-1 The Preliminary Operations Table

No.	Event	Changed signal	Conflicting outputs	Synthesis
1	A	b'	$A, \bar{A}$	
2	B	a	$B, \bar{A}$	
3	$\bar{B}$	b	$\bar{B}, A$	
4	$\bar{A}$	b'		
5	B	a'	$B, \bar{B}$	
6	A	b		
7	$\bar{A}$	a		
8	$\bar{B}$	a'		

conflicting outputs associated with the particular changed input are listed. For example, the events associated with the changed signal b' (row 1) can be either A or also $\bar{A}$ (row 4), and hence these different events are listed in the fourth column.

The synthesis is initiated by inspecting the conflicting outputs (or events) that prevail in the table. These contradictions are eliminated by the use of memory elements (Y's), which are assigned to each conflicting row in a Gray Code fashion; thus, an N-row contradiction will require M-memory elements where M is the next larger integer than $\log_2 (N)$. In the two-cylinder example, there are two contradictions relative to the changed input signal b' (rows 1 and 4); therefore, only one memory element is required for the elimination of the particular contradiction, which in Table 8-2 is indicated by Y_1. As can be seen in the fourth column of this table, y'_1 is inserted in the first row, while y_1 is placed in the fourth row. Thus, the contradiction between the two rows is eliminated. Similarly, Y_2, Y_3, Y_4 memory elements are utilized to eliminate the conflicting outputs relating to inputs a, b, and a', respectively.

The next step requires the inspection of the operations table for the needed auxiliary memory elements. These auxiliary elements are required when the input-memory combination obtained by the previous step is inadequate for the unique representation of each state (or row). Since this uniqueness requirement has already been satisfied in the example of Table 8-2, no auxiliary memory elements are required.

The assignment of the SET and RESET signals of the memory elements appears to be quite trivial, as no auxiliary Y-valves are utilized. Placing the SET and RESET signals of these valves in rows immediately preceding their required ON or OFF conditions will never create any conflicting conditions of the SET and RESET signals.

The augmentation of the W shut-off valves may require a more careful inspection. Shut-off valves are needed in order to terminate an output that prevails as a result of an input which is activated for a long period of time. For example, although b' is represented only in rows 1 and 4, it actually still exists in row 2. Thus, the output that causes event A in row 1 to occur still prevails in row 2. However, as the prevalence of this actuating signal will not affect the operation of the machine (due to the type of output circuit, see Figure 8-1), no shut-off valve is required. In contrast, consider the case represented in row 2. Here, signal ay'_2, which prevails during the three rows 2, 3, and 4, should cause cylinder B to extend. However, signal by'_3 in row 3 indicates that cylinder B should retract. Therefore, contradictory outputs occur at row 3, which requires that the output relating to row 2 be shut off. This is performed by assigning w_1 at row 2 while assigning w'_1 at row 3, as shown in Table 8-2. Note that a similar case also occurs between rows 6 and 7, which requires the augmentation of the second W-valve, as seen in the twelfth column of Table 8-2.

The final task is to determine the SET and RESET signals of the W shut-off valves. In this particular example, the RESET signals of the W-valves have to be placed in the rows where the W's should be OFF, due to the row adjacency of the conflicting outputs. As there are no auxiliary memory elements, the selection of the

Table 8-2 The Synthesized Operations Table

	System analysis			Memory augmentation			SET/RESET of memories				Shut-off augmentation	Shut-off SET/RESET	
No.	E	C/S	C/O	Y	Y_{dr}	Y_{aux}	Y_1	Y_2	Y_3	Y_4	W	W_1	W_2
1	A	b'	$A,\bar{A}$	y'_1				R			w'_1	S	
2	B	a	$B,\bar{A}$	y'_2			S		R		w'_1	R	
3	$\bar{B}$	b	$\bar{B},A$	y'_3									
4	$\bar{A}$	b'		y'_4									
5	B	a'	$B,\bar{B}$	y_3						R			S
6	A	b		y_2				S	S		w'_2		
7	$\bar{A}$	a									w'_2		
8	$\bar{B}$	a'		y_4			R			S			R

E = event
C/S = changed signal
C/O = conflicting outputs
Y_{aux} = auxiliary memory elements
Y_{dr} = "drafted" memory elements

239

SET signals should be made such that there are no conflicts with their associated RESET signals. Placing the SET signals of these valves in the rows immediately preceding their required ON conditions shows that the SET and RESET signals of each W-valve will never oppose each other.

In order to gain a better understanding of the change signal synthesis procedure, a more extensive problem is considered. The Operations Table for this twenty-four-event example is shown in Table 8-3. The problem is particularly appropriate, as it demonstrates the various conditions which must be considered in the memory and W-valve augmentation; therefore, it can provide a clear insight as to the various steps involved in the procedure. For example, the changed input a' is associated with three distinct outputs—A, B, and C. Therefore, according to the associated rule, two memories are required for the representation of the input a' without the existence of any conflicts in the outputs. Hence, $y'_1 y'_2$ is used to represent rows 1 and 16, which are associated with event A. However, the uniqueness requirements of each state have not been achieved, which requires that an auxiliary memory element be utilized for the differentiation of row 1 from row 16.

The selection of the appropriate points for the SET and RESET of the Y-memory elements needs more attention. Such selection should avoid the utilization of auxiliary memory elements. As indicated in the outline, the selection must first consider the "unavoidable" or "favorable" SET and RESET positions. For example, the SET signal of Y_1 is in an unavoidable location, as only one row exists between the y_1 and the y'_1 augmentation. On the other hand, the RESET signal for Y_1 at row 24 is in a favorable position, as it immediately precedes a y'_1 augmentation and as it does not involve an auxiliary memory element. The selection of row 17 as the SET point of Y_1 requires that row 15 also be selected in order to avoid the promotion of Y_{15} as a mandatory memory element. However, Y_1 is required to be "OFF" at row 16, and setting Y_1 at row 15 would definitely impair the required logic of the system. Therefore, memory element Y_{15} must be "drafted" as mandatory in order to avoid this contradictory condition. This promotion makes the selection of row 15 as the RESET location for Y_2 favorable. The selection of the RESET point of Y_3 and the SET point of Y_{15} needs more attention. Trying to place the RESET of Y_3 in row 1 only would require the promotion of Y_{17} as a mandatory memory element. However, further investigation reveals that the selection of row 16 as a complementary RESET point of Y_3 does not create any conflicting conditions. This multiple selection will eliminate the influence of Y_{17}; hence, its promotion would not be necessary. A similar situation occurs in the selection of the SET point of Y_{15}, which requires that row 1 also be selected as a complementary SET point of Y_{15}. Further inspection of the Operations Table shows that the SET signals for Y_7 and Y_8 remain undetermined due to the unselected auxiliary memory element Y_{18} (see column 17) in rows 13 and 22. Moving the SET position up one row causes no persistent contradictions and shows the redundancy of the memory element Y_{18}.

The final task in the synthesis is the assignment of the W's and their associated SET and RESET signals. It may be noticed that the SET/RESET assignment for W_1 is performed in the conventional manner used in Table 8-2. For W_2, the case is

Table 8-3 The Synthesized Operations Table of a Twenty-Four-Event Sequential Logic System

System analysis			Memory augmentation			SET and RESET of the memory valves																Shut-off augmentation	SET/RESET of W-valves				
No.	E	C/S	C/O	Y	Y_{dr}	Y_{aux}	Y_1	Y_2	Y_3	Y_4	Y_5	Y_6	Y_7	Y_8	Y_9	Y_{10}	Y_{11}	Y_{12}	Y_{13}	Y_{14}	Y_{15}	Y_{16}	W	W_1	W_2	W_3	W_4	W_5
1	A	a'	A,B,C	$y_1'y_2'$		y_{17}'			R													S	w_1'	R				
2	B	a	$\bar{A},\bar{B},B$	$y_3'y_4'$							R	R																
3	$\bar{A}$	b	$\bar{A},\bar{B},\bar{C}$	$y_5'y_6'$				S																					
4	C	a'		$y_1'y_2'$									R	R									w_2'		R				
5	$\bar{B}$	c	$A,\bar{B},\bar{C}$	$y_7'y_8'$											R	R										S			
6	$\bar{C}$	b'	$B,C,\bar{C}$	$y_9'y_{10}'$								S					R	R					w_2'			R			
7	B	c'	$A,\bar{A},B$	$y_{11}'y_{12}'$														R	R				w_3'						
8	$\bar{B}$	b		$y_5'y_6$	$y_{13}'y_{14}'$	$y_{13}'y_{14}'$										S				S		R	w_3'						
9	B	b'		$y_9'y_{10}$	y_{16}'	y_{16}'																S							
10	$\bar{B}$	b		$y_5'y_6$	$y_{13}'y_{14}$	$y_{13}'y_{14}'$																	w_3'						
11	B	b'		$y_9'y_{10}$	y_{16}	y_{16}'															R		w_3'						
12	$\bar{B}$	b		$y_5'y_6$	y_{13}	y_{13}'							S	S							S		w_5'						
13	C	b'		y_9		y_{18}'																	w_4'						
14	A	c		$y_7'y_8$						S											R		w_4'						
15	$\bar{A}$	a		$y_3'y_4$	y_{15}'	y_{15}'		R													S			R			R		
16	A	a'		$y_1'y_2$		y_{17}			R														w_4'						
17	$\bar{A}$	a		$y_3'y_4$	y_{15}	y_{15}'	S																						
18	B	a'		y_1						S																			
19	$\bar{C}$	b		y_5							S																		
20	A	c'		$y_{11}'y_{12}$									S																
21	$\bar{B}$	a		y_3														S					w_5'	S				S	
22	C	b'		y_9																	S		w_5'						
23	$\bar{C}$	c		y_7																			w_1'						
24	$\bar{A}$	c'		y_{11}			R			R																			R

241

considerably different due to the fact that the two conflicting events C (row 4) and $\bar{C}$ (row 6) are separated by only one row; hence, the resetting of W_2 must occur within that row. This assignment is needed in order to avoid the unnecessary delay that might occur due to the "shutting off" of the associated W-valve. The SET signal selection poses still a different problem. If the selection of the SET signal is made at state 3, which has the input-memory combination $by'_5y'_6$, it can be observed that this signal still prevails at state 5, thus opposing the intended RESET signal. In order to eliminate this opposing signal, the SET signal for W_2 is moved upward one row, the signal of which is terminated in state 4. A similar situation occurs in the SET signal selection of W_4.

Still another case is exhibited by the shut-off valve W_5, where it is represented twice in the table. This occurrence is due to event C relative to the input-memory signal $b'y_9$, which is at states 13 and 22. Note that this signal actually prevails at state 23, thus creating conflicts between the extension of cylinder C (event C) and its retraction (event $\bar{C}$).

Once the Operations Table is completed, the derivation of the network equations becomes quite trivial. For example, in relation to the first example (Table 8-2), event A is scheduled to occur at states 1 and 6; thus the actuating output signal is:

$$A = (1) + (6)$$
$$= b'y'_1 + by_3w_2 \tag{8-1}$$

Similarly, the output equations for the remaining events are:

$$A' = (4) + (7) = b'y_1 + ay_2 \tag{8-2}$$
$$B = (2) + (5) = ay'_2w_1 + a'y'_4 \tag{8-3}$$
$$B' = (3) + (8) = by'_3 + a'y_4 \tag{8-4}$$

The SET and RESET signals for the Y- and W-valves are determined by the input signals and the augmented mandatory Y-memory signals; thus:

$$\text{SET}_{W_1} = b'y'_1 \qquad \text{RESET}_{W_1} = by'_3 \tag{8-5}$$
$$\text{SET}_{W_2} = a'y'_4 \qquad \text{RESET}_{W_2} = ay_2 \tag{8-6}$$
$$\text{SET}_{Y_1} = by'_3 \qquad \text{RESET}_{Y_1} = a'y_4 \tag{8-7}$$
$$\text{SET}_{Y_2} = by_3 \qquad \text{RESET}_{Y_2} = b'y'_1 \tag{8-8}$$
$$\text{SET}_{Y_3} = a'y'_4 \qquad \text{RESET}_{Y_3} = ay'_2 \tag{8-9}$$
$$\text{SET}_{Y_4} = ay_2 \qquad \text{RESET}_{Y_4} = b'y_1 \tag{8-10}$$

The implementation of the above equations can be achieved by applying any of the methods discussed in Chapter 6. However, when three- or four-way spool-type valves are utilized or where a prepared path network is needed, all OR functions should be satisfied by using detented three-way valves and setting and resetting the

SEQUENTIAL LOGIC DESIGN—NONCLASSICAL

valve prior to the occurrence of the associated event. For example, the representation for event A (Equation 8-1) involves one OR function between state 1 and state 6. This can be attained by using a detented three-way valve, which RESETs at state 5 and SETs at state 8. All other outputs can be implemented in a similar fashion. The resulting prepared path network can be observed in Figure 8-2.

It was mentioned earlier that the Change Signal Method could be utilized for networks exhibiting different types of input and output circuits provided that some minor changes are made within the procedure. To illustrate this situation, consider the case where an input/output circuit requires the activation of its outputs according to the LSC shown in Figure 8-3. Note that it is now OUTPUTS that are considered in the specification and not EVENTS as it was assumed previously. The Operations Table is constructed in a similar manner by only replacing the term "event" with "outputs." The conflicting outputs are listed in groups with each group reflecting the total output state. When two output states are different, they are considered as conflicting. The outputs can be represented in two ways—either conventionally in state-by-state form or by using persisting input-memory combinations involving W-valves only when necessary. The latter approach is considered more advantageous as it minimizes the W-valves in the network. This approach is illustrated in the example shown in Table 8-4.

In Table 8-4, output P at state 1 is activated, while in its previous state (state 6) it was OFF. By letting P be activated in state 1, the corresponding signal $(a'y_1')$ remains activated until the next state, providing Y_1 is SET in state 3. Since output

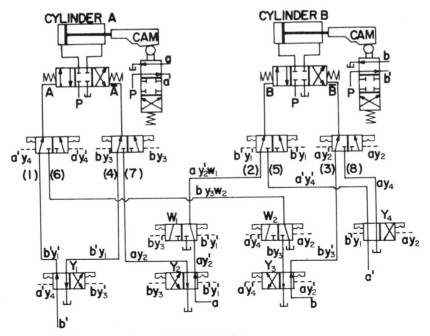

Figure 8-2 Prepared-path circuit implementation.

Outputs		Inputs (a, b)			
P	Q	00	01	11	10
1	0	(1)	2		
1	1		(2)	3	
0	1		4	(3)	
0	0		(4)	5	
1	1			(5)	6
0	1	1			(6)

Figure 8-3 The Logic Specification Chart.

Q is activated in state 2, the signal at this state can be utilized to perform the Y_1 setting task. Note, however, that since this signal remains activated until state 5, a W-valve is required to shut the signal OFF at state 4. The completed Operations Table can be observed in Table 8-4.

The SET and RESET signal assignments are made in a fashion similar to that presented in the procedure. Some minor difficulties tend to arise when setting and resetting W-valves. For example, in Table 8-4, although there is one row in between the row where W_1 should be ON and OFF, the resetting of W_1 is not performed at state 3, as W_1 has been utilized to control the activation of Q. The reader may also note that the RESET signal for Y_2 has been "moved" to state 1 to avoid the conflict with its SET signal.

In the formulation of the output equations, all the previous considerations should be carefully examined. Particularly, the excitation of the W- and Y-valves must be carefully analyzed, in order to know the actual activation of each output due to each input-memory-W combination reflected at each row. A careful selection of the input-Y-W combination should result in a minimal-element network. For example, in the case of output P in Table 8-4, only states 1 and 5 should be considered, since the output signal generated by the input-memory combination for state 1 remains activated during state 2. By applying the rationale given above, the

Table 8-4 The Completed Operations Table

	System analysis			Memory augmentation			Memory SET/RESET		Shut-off augmentation	Shut-off SET/RESET		
No.	Outputs (P, Q)	C/S	C/O	Y	Y_{dr}	Y_{aux}	Y_1	Y_2	W	W_1	W_2	W_3
1	1 0	a'	PQ'	Y'_1				R	w'_3	S		R
2	1 1	b	$PQ, P'Q'$						w_1			
3	0 1	a	$P'Q, PQ$	y'_2			S					
4	0 0	a'		y_1				S	w'_1	R	S	S
5	1 1	a		y_2					w_2, w_3			
6	0 1	b'	$P'Q$					R	w'_2		R	

following system equations can be derived:

$$P = a'y_1' + ay_2w_2 \qquad (8\text{-}11)$$
$$Q = bw_1 + ay_2w_3 \qquad (8\text{-}12)$$
$$\text{SET}_{W_1} = \text{RESET}_{W_3} = \text{RESET}_{Y_2} = a'y_1' \qquad (8\text{-}13)$$
$$\text{SET}_{Y_1} = ay_2' \qquad (8\text{-}14)$$
$$\text{RESET}_{W_1} = \text{SET}_{W_2} = \text{SET}_{W_3} = \text{SET}_{Y_2} = a'y_1 \qquad (8\text{-}15)$$
$$\text{RESET}_{W_2} = \text{RESET}_{Y_1} = b' \qquad (8\text{-}16)$$

In conclusion, it can be said that the method is applicable to most deterministic-type networks. It has been established that the method is most efficient for solving a particular type of feedback network. In the case of other types of networks, the method may yield unwanted, complex networks due to the prevalence of W-valves. It has been demonstrated, however, that the method is quite simple in comparison to the classical synthesis. Also since the synthesis can be performed on the single table, it offers a concise view not given by the other techniques. The user of the method should be extremely careful that he does not overlook conflicting signals. A formal procedure is needed to mechanize the search for these signals.

8-3 THE TOTAL SIGNAL METHOD

The concept of using a "tabular"-type logic specification as the basis for a stochastic (also known as irregularly activated) network-synthesis method was advanced by G. E. Maroney in 1969. He recognized that a network having random inputs could not be represented without considering the complete input state. Thus, the technique which he developed for synthesizing large-scale stochastic networks has become appropriately known as the Total Signal Method. Even though the method uses the total input signals in representing the system, the application of Boolean simplification techniques allows the derivation of network equations having near-minimal form.

In comparison with the classical synthesis, the method exhibits advantageous aspects similar to those of the Change Signal Method:

1 The method is quite simple to use. All synthesis operations, except for some algebraic simplification of the final results, can be performed on the synthesis table.
2 The procedure is capable of solving relatively large, finite-state machines involving large numbers of input variables.
3 Like the classical method, it is capable of solving irregularly activated (stochastic) networks.

However, there are certain drawbacks associated with the method as it now exists, in that some trial-and-error procedures are needed when a simplified network is required.

The Total Signal Method utilizes a tabular-type logic specification which is referred to as the Synthesis Table. This table was presented in Chapter 5, together with the rules for its construction. Synthesis by this method can be accomplished by applying the steps outlined below:

1. Form the preliminary Synthesis Table.
2. Inspect the table for nonunique input states. Count the number of times a particular input state occurs within the table.
3. Augment the inputs with memory elements to create uniqueness of all input states. Memory augmentation is performed in Gray Code fashion. If the number of memory states that can be achieved by the memories exceeds the number of nonunique input states, assign an additional memory state to the last nonunique input state in order to form memory states having the least number of memory variables.
4. For each augmented memory element, determine the necessary SET and RESET conditions required at each state of the machine. This is performed by placing a 0 for each state where Y' is needed as well as for each of the preceding states and placing a 1 for each state where Y is required and all its preceding states. All remaining states are assigned a "don't care" ("-").
5. SET and RESET signals are established by direct inspection of the table. In particular, locate the 0-states of each memory element that have an entry other than a 0-state in one or more of its previous states. Assign the 0-state an R or RESET state if a condition other than 0 exists. Similarly, locate each 1-state having one or more states other than 1 in its previous states. Such 1-states become the S or SET states for the particular memory elements.
6. Determine the output equations by utilizing the unique signal representation of each machine state associated with the "ON" output conditions. The output equations are represented by the total input signal plus the augmented memory signals of each row. The output expressions can be simplified using Karnaugh maps or other simplification procedures, considering all states as "don't cares" except the specified 0-states and 1-states.
7. Determine the SET and RESET equations. For the SET expression, all S's are considered as 1's, R's and zeros are considered as the zeros of the SET expression, while the 1's, the unused states, and specified "don't cares" are indicated as "don't cares." Similarly, for the RESET expression, all R's are the necessary RESET signals, while the S's and 1's are considered as the zeros of the RESET expression.

In order to illustrate the Total Signal Method, consider the example problem illustrated by the preliminary Synthesis Table shown in Table 8-5. Note that in the fourth column, the previous possible states are also listed. An inspection of the table reveals that input states $a'b'$ and ab' are not unique, and therefore memory elements will be required to establish the uniqueness of each state. As each of these states occurs exactly twice in the table, only one memory element is required to satisfy each input state. The assignment of the memories is shown in the sixth column of Table 8-6.

The next step in the synthesis is to determine the excitation conditions of the memory elements. As in the Change Signal Method, this method requires that the

SEQUENTIAL LOGIC DESIGN—NONCLASSICAL   247

Table 8-5 The Preliminary Synthesis Table

No.	Outputs $(Z_1\ Z_2)$		Input state	Next state	Previous states	Synthesis
1	1	0	$a'b'$	2, 3	3, 6	
2	1	0	ab'	4, 5	1, 4	
3	1	1	$a'b$	1	1, 4	
4	1	1	$a'b'$	2, 3	2	
5	0	0	ab	6	2	
6	0	0	ab'	1	5	

memories be SET or RESET prior to their use in establishing uniqueness of the input states. Therefore, the desired condition of the memory should be assigned for all previous possible states of the particular memory-augmentation state. Such conditions should be reflected in the synthesis table in order to assist in the selection of the optimal SET and RESET locations as illustrated in columns 7 and 8 of Table 8-6. For example, it can be observed that the Y_1 memory is utilized in rows 1 and 4. As the y_1 augmentation occurs at row 4, it is obvious that this condition is required in that row, and hence, a 1 is entered in row 4 in the Y_1 column. Further examination of the table reveals that there is only one possible previous state, which is row 2. Consequently, the condition 1 must be entered in this state. Similarly, for the y'_1 augmentation, rows 1, 3, and 6 should contain 0's.

The determination of the SET and RESET signals is performed in almost the same manner as in the Change Signal Method. For the RESET signal, every state previous to a 0-condition state is examined to determine whether they have an entry other than a 0. Every time one such case is detected, the earlier mentioned 0-condition is selected as the RESET state, and an R is entered in the particular row location in a new column (Y^* column). For example, in the Y_1^* column of Table 8-5, R's are entered in the third and sixth rows, since the previous states to these rows, 4 and 5, have entries other than 0's. Similarly, an S is inserted in row 2 because of the 0-condition in one of its previous states, row 1. The Y^* column is completed by transferring all other entries to their respective empty row locations.

Table 8-6 The Completed Synthesis Table

No.	System analysis					Memory augmentation				
	Outputs $(Z_1\ Z_2)$		Input state	Next state	Previous states	Memory	Y_1	Y_2	Y_1^*	Y_2^*
1	1	0	$a'b'$	2, 3	3, 6	y'_1	0	0	0	R
2	1	0	ab'	4, 5	1, 4	y'_2	1	0	S	0
3	1	1	$a'b$	1	1, 4		0	—	R	—
4	1	1	$a'b'$	2, 3	2	y_1	1	0	1	0
5	0	0	ab	6	2		—	1	—	S
6	0	0	ab'	1	5	y_2	0	1	R	1

The circuit equations can now be derived by direct interpretation of the synthesis table. Simplification of the equations can be achieved by applying map or tabular techniques to the resulting equation. As there are no unused input states in the logic specification, the expressions for the outputs are simply obtained by considering all the 1's in the table, i.e.:

$$Z_1 = a'b'y_1' + ab'y_2' + a'b + a'b'y_1 \qquad (8\text{-}17)$$
$$Z_2 = a'b + a'b'y_1 \qquad (8\text{-}18)$$

By simple Boolean manipulations or by inspecting the Karnaugh maps shown in Figure 8-4, it is apparent that the equations can be simplified to yield:

$$Z_1 = a' + b'y_2' \qquad (8\text{-}19)$$
$$Z_2 = a'b + a'y_1 \qquad (8\text{-}20)$$

The SET and RESET equations for the memories are obtained by a similar manner, only the "don't cares" are now available. The 1 terms for the RESET expressions are the R terms, while the 0's and the "–"s are used as "don't cares." Similarly, the SET expressions are constructed by the S's and simplified using the 1's and "–"s as "don't cares." Thus, the memory expressions are as follows:

$$\text{SET}_{Y_1} = ab'y_2' + (a'b'y_1 + ab) \qquad (8\text{-}21)$$
$$\text{RESET}_{Y_1} = a'b + ab'y_2 + (a'b'y_1' + ab) \qquad (8\text{-}22)$$
$$\text{SET}_{Y_2} = ab + (a'b + ab'y_2) \qquad (8\text{-}23)$$
$$\text{RESET}_{Y_2} = a'b'y_1' + (ab'y_2' + a'b + a'b'y_1) \qquad (8\text{-}24)$$

The Karnaugh maps (Figure 8-5) for these expressions reveal the following simplified solutions:

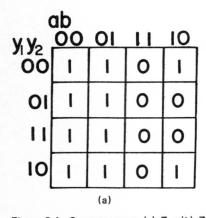
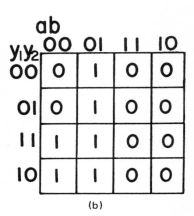

Figure 8-4 Outputs maps: (a) Z_1; (b) Z_2.

SEQUENTIAL LOGIC DESIGN—NONCLASSICAL

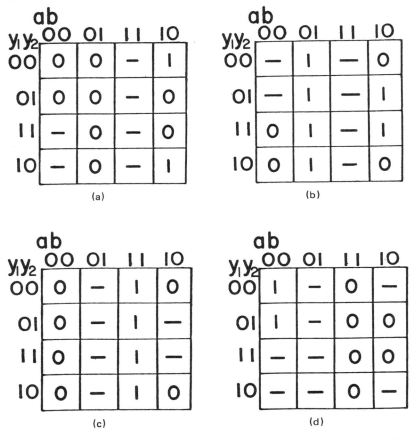

Figure 8-5 Karnaugh maps for the SET and RESET equations: (a) SET Y_1; (b) RESET Y_1; (c) SET Y_2; (d) RESET Y_2.

$$\text{SET}_{Y_1} = ay_2' \qquad \text{RESET}_{Y_1} = b + ay_2 \qquad (8\text{-}25)$$
$$\text{SET}_{Y_2} = b \qquad \text{RESET}_{Y_2} = a' \qquad (8\text{-}26)$$

The network can be implemented as shown in Figure 8-6.

In relation to multiple-output simplification, it may prove important to consider the "unsimplified" terms that are conjoined with a memory state in order to use each memory element as a passive device. For example, if the equations for the above circuit uses memory state Y_2 three times, then leaving these equations in the form $ab'y_2$ or $ab'y_2'$ may result in a further simplification of the network. Applying this technique for the example problem results in the network configuration given in Figure 8-7.

So far, trial-and-error methods in the synthesis have been avoided. As was mentioned earlier, trial-and-error approaches may be needed in order to further simplify the results. This is performed by trying every combination of 1's and 0's in

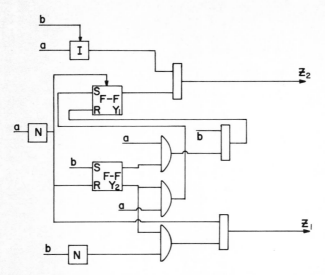

Figure 8-6 Network construction by individual output synthesis.

the "-" locations of the Y-columns. For example, a simplification can be obtained by inserting a 0 in the "-" location of Y_1, which alters the SET and RESET equations of Y_1 as follows:

$$\text{SET}_{Y_1} = ab'y_2' \qquad \text{RESET}_{Y_1} = b \tag{8-27}$$

which clearly demonstrates that further simplification was possible as shown in Figure 8-8.

8-4 THE STATE MATRIX METHOD

In 1970, R. L. Woods introduced a unique formatting procedure for recording and synthesizing a logic problem. Initially, the matrix was utilized for merely describing

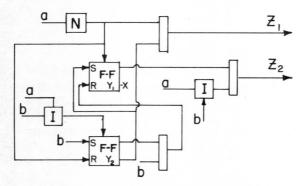

Figure 8-7 Network configuration by multiple output synthesis.

SEQUENTIAL LOGIC DESIGN—NONCLASSICAL

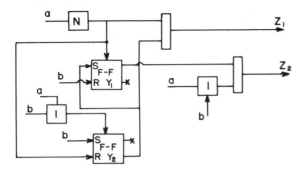

Figure 8-8 Network configuration by trial-and-error approaches.

the synthesis of both stochastic- and deterministic-type problems. As was noted in Chapter 5, the utilization of the matrix as a means for representing a logic specification has created a different view of logic synthesis.

Basically, the State Matrix Method, as it is now called, attempts to partition the existing states into groups of similar inputs and outputs in order to yield a unique insight into the input and output relationships. The synthesis procedure is very similar to that of the Total Signal Method discussed in the previous section (Woods also recognized an approach for deterministic problems that utilizes a variation of the Change Signal Method). As with the previously discussed methods, the State Matrix Method performs all synthesis operations within the matrix. The approach can be outlined as follows:

1. Form the preliminary State Matrix.
2. Inspect the matrix for nonunique states. Nonunique states are those that occur more than once in a column and at rows relating to the same output. Count the number of times such nonunique states occur in a particular column (or input state).
3. Augment with memory elements in order to create uniqueness for all machine states. Consider each input state individually and assign memories in a Gray Code fashion. If the number of memory states that can be achieved by the augmented memories exceeds the number of nonunique input states, assign more than one memory state to the last nonunique input state in order to form memory states having the fewest number of memory variables.
4. Determine the output equations. The output equations are represented by the states shown in the ON row of each output, together with their memory augmentation.
5. Determine the SET and RESET states for each memory element. As in the total signal approach, the setting and resetting operations have to be performed before the particular condition is required by its augmentation. This method selects all states that might have occurred prior to the state where augmentation takes place.
6. Determine the SET and RESET expressions. These are obtained by equating all the SET and RESET states as recorded in Step **4**.

$$\begin{bmatrix} Z_1 \\ Z_1' \\ Z_2 \\ Z_2' \end{bmatrix} = \begin{array}{|c|c|c|c|} \hline & & 8(1,4) & & \\ \hline 1(2,3) & 2(1,4) & \begin{array}{c}4(2,6)\\5(2,3)\\7(6,8)\end{array} & \begin{array}{c}3(1,5)\\6(1,7)\end{array} \\ \hline & & 5(2,3) & \\ \hline 1(2,3) & \begin{array}{c}2(1,4)\\8(1,4)\end{array} & \begin{array}{c}4(2,6)\\7(6,8)\end{array} & \begin{array}{c}3(1,5)\\6(1,7)\end{array} \\ \hline \end{array} \quad \begin{bmatrix} 0 & 0 \\ 0 & 1 \\ 1 & 1 \\ 1 & 0 \end{bmatrix}$$

(Output matrix) — (Network matrix) — (Input matrix, $a\ b$)

Figure 8-9 Preliminary State Matrix.

7 Simplify the outputs, SET, and RESET equations by using Karnaugh maps or other simplification techniques. All unused input states may be considered as "don't cares."

In order to provide a clear understanding of the procedure, the example problem represented by the Preliminary State Matrix shown in Figure 8-9 is considered. Note that the second column, which relates to the input state ($ab = 01$) exhibits two nonunique machine states, which are states 2 and 8. In the third column, three nonunique machine states exist, while in the fourth column there are two states. The next states are exhibited by the entries in parentheses following each state. In accordance with Step 3, memory augmentation is necessary for all nonunique input states. Therefore, memory assignments are made in a similar fashion as the Total Signal Method, i.e., by utilizing the Gray Code assignment for each nonunique input state. The completed assignments can be observed in Figure 8-10. Note that the assignments made for one output (for example the first two rows) should be reproduced in the other rows such that no conflicts occur.

$$\begin{bmatrix} Z_1 \\ Z_1' \\ Z_2 \\ Z_2' \end{bmatrix} = \begin{array}{|c|c|c|c|} \hline & & 8(1,4)\ y_1' & & \\ \hline 1(2,3) & 2(1,4)\ y_1 & \begin{array}{c}4(2,6)\ y_2'y_3'\\5(2,3)\ y_2'y_3\\7(6,8)\ y_2\end{array} & \begin{array}{c}3(1,5)\ y_4'\\6(1,7)\ y_4\end{array} \\ \hline & & 5(2,3)\ y_2'y_3 & \\ \hline 1(2,3) & \begin{array}{c}2(1,4)\ y_1\\8(1,4)\ y_1'\end{array} & \begin{array}{c}4(2,6)\ y_2'y_3'\\7(6,8)\ y_2\end{array} & \begin{array}{c}3(1,5)\ y_4'\\6(1,7)\ y_4\end{array} \\ \hline \end{array} \quad \begin{bmatrix} 0 & 0 \\ 0 & 1 \\ 1 & 1 \\ 1 & 0 \end{bmatrix}$$

(Output matrix) — (Network matrix) — (Input matrix, $a\ b$)

Figure 8-10 Augmented State Matrix.

SEQUENTIAL LOGIC DESIGN—NONCLASSICAL   253

The output and SET/RESET equations can be derived directly from the table. For each of the outputs, all the states represented in the "ON" row are used to satisfy the outputs, along with their memory augmentations. Thus, for the example being considered, the output equations are:

$$Z_1 = 8 = a'by'_1 \tag{8-28}$$
$$Z_2 = 5 = aby'_2 y_3 \tag{8-29}$$

The memory equations are determined in a manner similar to the output equations. However, as the SET and RESET signals should occur prior to the augmentation of the particular memory condition, it is necessary to inspect the matrix for machine states that might occur prior to the particular augmentation. Note that this inspection need only be performed in one part of the matrix—the part that relates to one single-output function of the network. For example, considering the "submatrix" for output Z_1, it can be seen that machine states 1, 4, and 5 may possibly occur prior to state 2, where y_1 has been augmented. Thus, the augmented signals relating to all such states must be utilized as the necessary SET signals for the memory valve Y_1. Similarly, as the RESET condition of memory Y_2 is required by machine states 4 and 5, all states that might precede these two states should be selected as the RESET states for the memory Y_2. Inspecting all parenthetical entries in the matrix reveals that states 2, 3, and 8 are the RESET states for memory Y_2. Hence, the SET and RESET equations for the four memory elements are as follows:

$$\text{SET}_{Y_1} = 1 + 4 + 5 = a'b' + ab(y'_2 y'_3 + y'_2 y_3) = a'b' + aby'_2 \tag{8-30}$$
$$\text{RESET}_{Y_1} = 7 = aby_2 \tag{8-31}$$
$$\text{SET}_{Y_2} = 6 = ab'y_4 \tag{8-32}$$
$$\text{RESET}_{Y_2} = 2 + 3 + 8 = a'by_1 + ab'y'_4 + a'by'_1 = a'b + ab'y'_4 \tag{8-33}$$
$$\text{SET}_{Y_3} = 3 = ab'y'_4 \tag{8-34}$$
$$\text{RESET}_{Y_3} = 2 + 8 = a'b \tag{8-35}$$
$$\text{SET}_{Y_4} = 4 + 7 = aby'_2 y'_3 + aby_2 = aby'_3 + aby_2 \tag{8-36}$$
$$\text{RESET}_{Y_4} = 1 + 5 = a'b' + aby'_2 y_3 \tag{8-37}$$

The network configuration can be observed in Figure 8-11.

Because of the simplicity of the problem solution, the above equations have been simplified by using the postulates and theorems of Boolean algebra. Obviously, when more complex solutions are encountered, the use of Karnaugh maps or other simplification procedures cannot be avoided.

8-5 TRANSITION TABLE METHOD

In 1970, a new synthesis method utilizing the conventional Primitive Flow Table was introduced by R. M. H. Cheng and K. Foster. The method was based upon the presumption that most fluid(ic) type logic circuits are not complex, and hence,

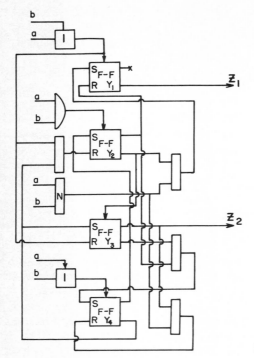

Figure 8-11 Network implementation.

minimal row-reduction approaches need not be performed. The method is particularly effective when used for synthesizing deterministic-type logic networks. It is also effective when a "small" degree of irregular conditions (stochastic) are involved in the logic system. Although it can basically handle almost any given problem, stochastic-type problems often create difficulties, which influence the degree of minimality of the network.

The Transition Table Method is suited to resolve control networks having an input-output circuit as shown in Figure 8-12. All outputs are expected to have a Flip-Flop-type memory element (a detented valve) which is SET and RESET in a manner that will satisfy a prescribed set of output conditions. The inputs are

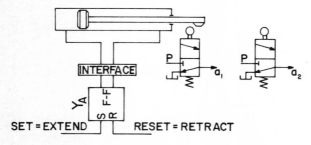

Figure 8-12 Input-output circuit.

classified into two categories: one category (which should be the major category) consists of feedback-type input signals which indicate the completion of specific operations of the outputs. Under the second category are the external inputs. Since these external inputs may occur at random, they are the sole contributors in the creation of a stochastic network. The number of feedback-type signals is exactly twice the number of cylinders in the system, while the remaining inputs are the external inputs.

As in the original Huffman technique, the logic description is recorded in the Primitive Flow Table. However, since the reader is now familiarized with the use of the LSC, this format will be used in this section rather than the PFT. The procedure for the Transition Table Method can be outlined as follows:

1 Construct the LSC from the logic specification.
2 Perform row "minimization" using the following criteria:
  a Only rows having identical outputs may be combined.
  b Two rows are conjoined if in each column the state numbers are the same or if one of the entries is undetermined or "-".
3 Relabel the stable states by using the row numbers. Hence, a row may contain more than one stable state having the same state number. This step is performed only for convenience in later steps. Relabel the unstable states using the corresponding stable-state numbers.
4 Using the output memory elements to represent each row, determine whether or not each row has been represented uniquely by these memory elements.
5 If the required uniqueness is not satisfied, assign new memory variables until all rows contain suitable memory states. This can be accomplished as follows:
  a Start forming the first group of unique rows by using the first row of the LSC as the first row of the group.
  b The second row of the LSC can be included in the group if it preserves the uniqueness condition of the group. Such a condition exists when the combination of inputs and secondary variables are different.
  c Add the third subsequent row of the LSC to the group if it also preserves the group uniqueness.
  d When a nonunique condition is encountered, start forming a new group in a manner similar to the first group.
  e After all rows of the LSC have been placed into proper groups, determine the number of memory elements needed to represent each of the groups and assign the memory states in appropriate Gray Code fashion.
6 Satisfy memory transition requirements by using the following technique:
  a Consider one stable state and one of its related unstable states. Record both memory (secondary) states of the stable and unstable states.
  b By changing one memory variable at a time, record all possible ways the transition may occur in order that the control system can reach the stable state under consideration.
  c Discard any "path" that involves a row that requires a transition to a different stable state.
  d Select the "path" that includes the least number of rows that have no transitional requirements (or undetermined, "-").
  e If no such path exists, add one memory variable and assign all existing rows

to the 0-condition, thus creating new, empty rows represented by the 1-condition of the particular memory variable. Return to step **6**.

f Represent the selected transitional path on the chart by replacing all the unstable states and "–"s with the required next state in the transition.

g Consider the next unstable state.

h Consider the next stable state.

7 Develop the excitation maps for each memory element (including those of the outputs) as performed in the Classical Synthesis.

8 Derive the SET and RESET equations.

The Transition Table Method can be demonstrated by means of an example. Consider a network for controlling two cylinders having the following sequence: A, B, $\bar{B}$, $\bar{A}$, B, A, $\bar{A}$, $\bar{B}$, where A and $\bar{A}$ are used to represent the extension and retraction of cylinder A. The input and output circuits are as shown in Figure 8-12. Inputs a_1 and a_2 are used to represent the retracted and extended positions of cylinder A, respectively, while b_1 and b_2 relate to similar conditions of cylinder B. The output memory elements are subscripted by the output they represent: Y_A for cylinder A, etc. The LSC of the particular problem can be observed in Figure 8-13. Note that the outputs are represented by Y_A and Y_B for practical considerations.

The second step requires the merging of rows having similar outputs and next states. For example, rows 1 and 2 can be merged, as they have the same outputs and no conflicting state numbers exist within any of the columns. Row 2 cannot be merged with row 6, as the next states do not agree with each other (3 and 7). One configuration resulting from this merging step can be observed in Figure 8-14. A

Y_B	Y_A	b_2=0 b_1=1 a_2=0 a_1=1	0 1 0 0	0 1 1 0	0 0 1 0	1 0 1 0	0 0 0 1	1 0 0 1	1 0 0 0
0	1	(1)	2						
0	1		(2)	3					
1	1			(3)	4				
1	1				(4)	5			
0	1				6	(5)			
0	1			7	(6)				
0	0	8	(7)						
0	0	9	(8)						
1	0	(9)							
1	0						10		
1	0						(10)	11	
1	1							(11)	12
1	1					13			(12)
1	0					(13)			14
1	0							15	(14)
0	0							16	(15)
0	0	1							(16)

Figure 8-13 Logic Specification Chart.

SEQUENTIAL LOGIC DESIGN—NONCLASSICAL

		b_2	0	0	0	0	1	0	1	1
		b_1	1	1	1	0	0	0	0	0
		a_2	0	0	1	1	1	0	0	0
Y_B	Y_A	a_1	1	0	0	0	0	1	1	0
0	1		(1)	(2)	3					
1	1				(3)	(4)	5			
0	1				7	(6)	(5)			
0	0		9	(8)	(7)					
1	0		(9)					(10)	11	
1	1						13		(11)	(12)
1	0						(13)		15	(14)
0	0		1					(16)	(15)	

Figure 8-14 "Merged" Specification Chart.

slight modification of the chart can be made by changing all the state numbers to agree with the number of the row, the result of which can be observed in Figure 8-15. Following this modification, an unstable state, e.g., unstable state 3, indicates that the secondary states must change to row 3 where its stable state can be found.

Inspection of Figure 8-15 shows that the row representation given by the memory elements of the outputs does not establish a unique representation of each row. Therefore, additional memory elements are required, which are assigned according to Step 5 of the procedure. This assignment is performed by forming a group that initially consists of only the first row, which has the output memory combination of (01). The second row has an output combination of (11); hence, it can be included within the first group. The third row, however, is represented by $Y_A Y_B = (01)$, which is already represented in group 1, and, therefore, a second group is established. By the same means, the grouping process is completed, which

Row no.	Group no.	Y_2	Y_1	Y_B	Y_A	b_2 0	0	0	0	1	0	1	1
						b_1 1	1	1	0	0	0	0	0
						a_2 0	0	1	1	1	0	0	0
						a_1 1	0	0	0	0	1	1	0
1	I	0	0	0	1	(1)	(1)	2					
2				1	1			(2)	(2)	3			
3				0	1				4	(3)	(3)		
4	II	0	1	0	0	5	(4)	(4)					
5				1	0	(5)					(5)	6	
6				1	1					7		(6)	(6)
7	III	1	1	1	0					(7)		8	(7)
8				0	0	1					(8)	(8)	

Figure 8-15 Modification and grouping in the MSC.

results in three groups, as shown in Figure 8-15. Two memories are required, which are assigned to each of the groups by a Gray Code representation.

The next step involves the adjustments of the required transitions. Multiple memory changes should be avoided; therefore, transitions that involve multiple memory changes must be "routed" through a path in the table such that only one memory changes at a time. Additional memory elements should be used only when necessary. For example, consider the transition between stable state 1 and its unstable state (at row 8). The memory state before transition (row 8) is $Y_A Y_B Y_1 Y_2 = 0011$, while the intended final state is 1000. As the transition involves the change of memories Y_A, Y_1, Y_2, the paths are determined by the order in which memories should be changed. Note that by the law of permutation a transition that involves N memory changes have $N!$ possible paths. Thus, for the transition from row 8 to row 1, the following paths are possible:

0011-1011-1010-1000;   0011-1011-1001-1000;
0011-0001-1001-1000;   0011-0001-0000-1000;
0011-0010-1010-1000;   0011-0010-0000-1000;

As the selections are of equal importance, the first path is selected at random. Similarly, the transitional requirements for states 3 and 7 may be observed as shown in Figure 8-16.

Once the transitions in the chart have been satisfied, the formation of the excitation map can be initiated. Using the same rules as given for the Classical Method, the SET (S) and RESET (R) positions for each of the memory elements can be determined. A "compacted" form of the excitation maps, which do not show the rows and columns that contain only "don't cares," can be observed in Figure 8-17. The SET and RESET equations can hence be derived by conventional means, resulting in the following set of equations:

Row no.	Group no.	Y_2	Y_1	Y_B	Y_A	b_2=0 b_1=1 a_2=0 a_1=1	b_2=0 b_1=1 a_2=0 a_1=0	b_2=0 b_1=1 a_2=1 a_1=0	b_2=0 b_1=0 a_2=1 a_1=0	b_2=1 b_1=0 a_2=1 a_1=0	b_2=0 b_1=0 a_2=0 a_1=1	b_2=1 b_1=0 a_2=0 a_1=1	b_2=1 b_1=0 a_2=0 a_1=0
1	I	0	0	0	1	(1)	(1)	2		3			
2				1	1			(2)	(2)	≴1			
3	II	0	1	0	1	1		4	(3)	(3)			
4				0	0	5	(4)	(4)					
5				1	0	(5)				7	(5)	6	
6				1	1					⁊5		(6)	(6)
7	III	1	1	1	0					(7)		8	(7)
8				0	0	⁊9					(8)	(8)	
9				0	1	3							

Figure 8-16 Adjustments of transitions.

SEQUENTIAL LOGIC DESIGN—NONCLASSICAL

Y_2	Y_1	Y_B	Y_A	b_2 0 b_1 1 a_2 0 a_1 1	0 1 0 0	0 1 1 0	0 0 1 0	1 0 1 0	0 0 0 1	1 0 0 1	1 0 0 0
0	0	0	1	0001	0001	00S1		0S01			
0	0	1	1			0011	0011	00R1			
0	1	0	1	0R01		010R	0101	0101			
0	1	0	0	01S0	0100	0100					
0	1	1	0	0110				S110	0110	011S	
0	1	1	1					011R		0111	0111
1	1	1	0					1110		11R0	1110
1	1	0	0	110S					1100	1100	
1	1	0	1	R101							

Figure 8-17 Excitation map.

$$\text{SET}_{Y_A} = a_1 b_1 y_2 + a_1 b_2 y_2' \qquad \text{RESET}_{Y_A} = a_2 b_1 y_1 + a_2 y_B y_1 \qquad (8\text{-}38)$$

$$\text{SET}_{Y_B} = a_2 b_1 y_1' + a_1 y_A' y_2' \qquad \text{RESET}_{Y_B} = a_1 y_2 + a_2 b_2 y_1' \qquad (8\text{-}39)$$

$$\text{SET}_{Y_1} = b_2 y_B' \qquad \text{RESET}_{Y_1} = a_1 b_1 y_A y_2' \qquad (8\text{-}40)$$

$$\text{SET}_{Y_2} = a_1' b_1' y_A' \qquad \text{RESET}_{Y_2} = y_A \qquad (8\text{-}41)$$

The network for this example can be observed in Figure 8-18. Note that there are no output equations, as they have been incorporated in the memory elements.

8-6 THE STATE DIAGRAM METHOD

The synthesis of logic circuits by means of diagramming was initially introduced by P. I. Chen and Y. H. Lee in 1974. Basically, the method is designed to resolve deterministic, feedback-type sequential logic networks which have input-output circuits characterized by the illustration in Figure 8-19. By graphical means, the method attempts to reduce the number of memory elements required by the network. Unlike the method devised by Cole, this approach utilizes the true-valued, total signal for the representation of the primary and secondary outputs. By the intuitive recognition of the excitation patterns of the inputs, outputs, and memory variables, it is possible to select the near minimal configuration of the circuit.

The procedure for the State Diagram Method can be outlined as follows:

1 Construct and augment the synthesis table (as shown in Table 8-7) in the following manner:
   a Place state numbers in the first column.
   b List all associated feedback signal combinations (considering only the activated, true-valued input signals) in the second column.
   c Ignore all intermittent states during which a cylinder is extending or retracting.

    d List the required control outputs in column 3.
    e Augment with memories all the nonunique Feedback signal combinations.
2 For each augmented memory element, list all possible SET and RESET combinations that can be used for satisfying the excitation requirements of the particular memory element. The selection of each of these SET and RESET signals should avoid the utilization of memory (secondary) signals. In particular, perform the selection as follows:
    a For each memory element search for all possible pairs of input states that can

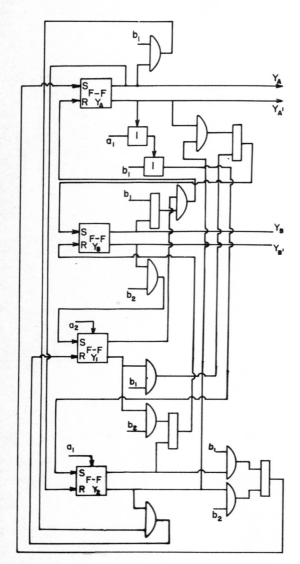

Figure 8-18 Network implementation.

SEQUENTIAL LOGIC DESIGN—NONCLASSICAL

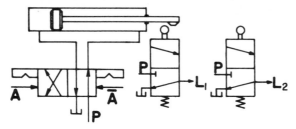

Figure 8-19 Input and output circuit.

be utilized to represent the SET and RESET signals of the particular memory. Continue to the next memory element if one or more such pairs exists.

  b If no unaugmented SET and RESET signals are found, then include one memory into each of the machine-state representations. If a state is augmented by more than one memory variable, try to include each of the memories individually. Again, if one or more pairs of SET/RESET signals are found, proceed to the next memory element.

  c If still no pair of SET and RESET signals are obtained, try two, three, etc., memory augmentations until each memory element has a set of minimal memory-augmented SET and RESET signals.

3 Construct the state diagram of the secondary variables for each of the SET and RESET conditions. The diagram should represent the actual condition of each memory element at any given time.

4 Compare the diagrams to determine whether or not there are similar excitation patterns of one memory to another. These memories are said to be equivalent, provided that the particular SET and RESET signals are used. Indicate this similarity by recording in parentheses the particular diagram by the memory denotation of the other diagrams (which can be in the complemented form if the excitation pattern is exactly opposite to each other).

5 By careful consideration of each diagram, select the SET and RESET signals such that the least number of memory variables are required to represent the problem.

6 Construct the state diagram for the selected memories, inputs, and the required outputs. The outputs that are not determined can be represented by dotted segments to indicate "don't cares."

Table 8-7 The Synthesis Table

State	Feedback signal combination	Operation	Secondary variable
1	$L_1 L_3$	A	Y_1
2	$L_2 L_3$	B	Y_2
3	$L_2 L_4$	$\bar{B}$	Y_3
4	$L_2 L_3$	$\bar{A}$	Y'_2
5	$L_1 L_3$	B	Y'_1
6	$L_1 L_4$	A	Y_4
7	$L_2 L_4$	$\bar{A}$	Y'_3
8	$L_1 L_4$	$\bar{B}$	Y'_4

7 Formulate the control outputs individually by searching for a minimal combination of the true-valued inputs and the secondary variables that can be utilized for representing the outputs. The "don't care" outputs should be effectively exploited in order to obtain a minimal output representation.

This graphical method can be illustrated by using the same two cylinder problem discussed in Section 8-2. The sequence of operation for the system is A, B, $\bar{B}$, $\bar{A}$, B, A, $\bar{A}$, $\bar{B}$. Note that in the synthesis table shown in Table 8-7, only true-valued feedback signals are considered, thus using L_1, L_3 instead of $L_1 L_2' L_3 L_4'$ etc. As there are four pairs of nonunique signal representations in the table, four memories are required to establish signal uniqueness. These memories are represented by Y_1, Y_2, Y_3, and Y_4 in column 4.

The next important task is the selection of the SET and RESET signals for the memory elements. By first considering the unaugmented feedback inputs, all possible pairs of signals that can satisfy the requirements of each memory element are listed as in Table 8-8. In connection with this particular problem, it is realized that the unaugmented input signals are sufficient for representing the memory elements; therefore, in order to optimize the use of each memory, no augmented signal need be utilized. The state diagram showing the excitation conditions of each memory element relating to each pair of SET and RESET signals is shown in Figure 8-20. The dotted lines of Figure 8-20 show the excitation of the memory, while the solid lines are the required SET or RESET conditions of the memory. Note that the memory should be SET or RESET in the state previous to the required SET or RESET locations.

Careful inspection of the diagram reveals the similarity of the pattern of the first row with that of the eighth row. Since the forms of the patterns are totally opposite to that of the other, the equivalences of the memories are also opposite,

Table 8-8 Potential SET and RESET Equations

Memory element	SET	RESET
Y_1	$L_1 L_4$ $L_2 L_4$ $L_1 L_4$	$L_2 L_3$ $L_2 L_3$ $L_2 L_4$
Y_2	$L_1 L_4$ $L_1 L_3$	$L_2 L_4$ $L_2 L_4$
Y_3	$L_2 L_3$ $L_1 L_3$ $L_2 L_3$	$L_1 L_3$ $L_1 L_4$ $L_1 L_4$
Y_4	$L_2 L_3$ $L_1 L_3$	$L_2 L_4$ $L_2 L_4$

SEQUENTIAL LOGIC DESIGN—NONCLASSICAL

SECONDARY VARIABLE	SET RESET	STATES 0 1 2 3 4 5 6 7
$Y(Y_3')$	$L_1 L_4\ L_2 L_3$	
$Y(Y_4')$	$L_2 L_4\ L_2 L_3$	
$Y_1(Y_2)$	$L_1 L_4\ L_2 L_4$	
$Y_2(Y_1)$	$L_1 L_4\ L_2 L_4$	
$Y_2(Y_4)$	$L_1 L_3\ L_2 L_4$	
Y_3	$L_2 L_3\ L_1 L_3$	
Y_3	$L_1 L_3\ L_1 L_4$	
$Y_3(Y_1')$	$L_2 L_3\ L_1 L_4$	
$Y_4(Y_1')$	$L_2 L_3\ L_2 L_4$	
$Y_4(Y_2)$	$L_1 L_3\ L_2 L_4$	

Figure 8-20 Memory state diagram for all possible SET and RESET functions.

i.e., Y_1 is equivalent to Y_3'. Similarly, it can be seen that Y_1 is equivalent to both Y_4' and Y_2. These similarities are recorded in the first column of Figure 8-20 (shown by the parenthetical entries). With this information, it is now possible to select the minimal number of memory elements that can be used to represent the logic system. It can be seen that by selecting the first (Y_1 and Y_3') and the fifth (Y_2 and Y_4) rows, all memory requirements have been satisfied.

Once the necessary memory elements have been selected, the complete state diagram of inputs, outputs, and memory elements can be constructed as shown in Figure 8-21. Note that the intermittent states are shown as shaded columns in the diagram. These columns can be excluded from the synthesis as long as only true-valued inputs are utilized for the representation of the outputs. In relation to this, the reader should note that all inputs are deactivated during these intermittent states, and hence, all complements of the inputs are ON during these periods. Utilizing the complements of the inputs without knowledge of such intermittent states may result in a premature activation of the outputs. In Figure 8-21, the determined outputs are shown by solid lines, while the don't care conditions are shown by dotted lines at both the ON and OFF conditions.

The output equations are determined by merely inspecting the diagram. This is performed by trying to obtain the minimal combination of signals needed to represent the particular output. For example, the outputs of the system under consideration are:

$$A = Y_1 Y_2 \tag{8-42}$$
$$\bar{A} = L_3 Y_2' + Y_1 Y_2' \tag{8-43}$$
$$B = Y_1' Y_2 \tag{8-44}$$
$$\bar{B} = Y_1' Y_2' + L_1 Y_2' \tag{8-45}$$

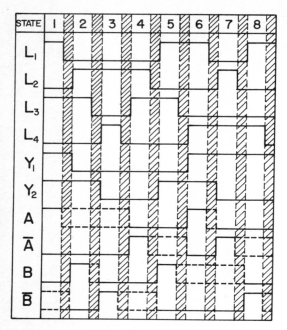

Figure 8-21 State diagram.

The implementation of these equations and the previously selected SET and RESET equations can be represented by the twelve-element network shown in Figure 8-22.

In the above example, the SET and RESET equations for all memory elements do not require any secondary augmentation. This may have given the impression that the procedure is quite simple. This is not always true, since the approach

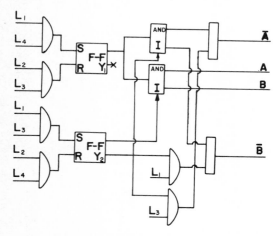

Figure 8-22 Network implementation.

SEQUENTIAL LOGIC DESIGN—NONCLASSICAL

Table 8-9 The Synthesis Table

State	Feedback signal combination	Operation	Secondary variable
1	$L_1 L_3$	A	$Y_1 Y_2$
2	$L_2 L_3$	B	$Y_3 Y_4$
3	$L_2 L_4$	$\bar{B}$	Y_5
4	$L_2 L_3$	$\bar{A}$	$Y'_3 Y_4$
5	$L_1 L_3$	B	$Y'_1 Y_2$
6	$L_1 L_4$	$\bar{B}$	Y_6
7	$L_1 L_3$	A	Y'_2
8	$L_2 L_3$	B	Y'_4
9	$L_2 L_4$	$\bar{A}$	Y'_5
10	$L_1 L_4$	$\bar{B}$	Y'_6

becomes more complex when any of the SET and RESET equations require the augmentation of one or more memory elements. In order to illustrate this, consider a two-cylinder system involving the following sequence: A, B, $\bar{B}$, $\bar{A}$, B, $\bar{B}$, A, B, $\bar{A}$, B. The synthesis table can be constructed as shown in Table 8-9. Six memory variables are required for the unique representation of each machine state. By utilizing all unaugmented feedback inputs, it can be observed that only memories Y_1, Y_3, and Y_6 can be satisfactorily SET and RESET using such signals. Therefore, it is necessary to involve one memory variable for the representation of each machine state. All combinations of SET and RESET (including combinations having augmented and unaugmented SET and RESET) must be checked and listed if they are applicable for a particular memory element.

The complete listing of SET and RESET signals is shown in Table 8-10. Note that every combination of inputs and memories is inspected for their applicability for the particular memory device. As all memories have an associated SET and RESET, no attempt is made to include dual memory augmentation.

The reader may realize that the establishment of equivalence may be directly conducted in the SET and RESET table of Table 8-10. This is accomplished very simply by comparing the SET and RESET equations of one memory to those of another. For example, the second SET/RESET function of Y_1 is $L_2 L_4$ and $L_2 L_3$, respectively, which is definitely the complement of the third SET/RESET function of Y_6. Thus, a Y'_6 is inserted in the second row of the Y_1 table; and a Y'_1 is inserted in the third row of the Y_6 table. Further inspection reveals that only three memory elements are required for representing the control network, which are $Y_1(Y'_6)$, $Y_2(Y_4, Y_5)$ and Y_3. The completed state diagram (without the intermittent states) is shown in Figure 8-23, which reveals the following output equations:

$$A = L_1 L_3 Y_3 \qquad (8\text{-}46)$$
$$\bar{A} = L_3 Y'_3 + L_4 Y_1 \qquad (8\text{-}47)$$

$$B = L_2L_3Y_3 + L_1L_3Y_3' \tag{8-48}$$
$$\bar{B} = L_4Y_2 + L_1L_4 \tag{8-49}$$

The SET and RESET equations are:

$$\text{SET}_{Y_1} = L_2L_4 \qquad \text{RESET}_{Y_1} = L_2L_3 \tag{8-50}$$
$$\text{SET}_{Y_2} = L_1L_4Y_1 \qquad \text{RESET}_{Y_2} = L_1L_4Y_1' \tag{8-51}$$
$$\text{SET}_{Y_3} = L_1L_4 \qquad \text{RESET}_{Y_3} = L_2L_4 \tag{8-52}$$

Note that Y_1' has been inserted for replacing the Y_6 representation in the SET and RESET equations of Y_2. The reader should realize that due to the nature of the synthesis approach, the method is limited to relatively small problems.

Table 8-10 Potential SET and RESET Equations

Equivalent to	SET	RESET	Equivalent to	SET	RESET
SET/RESET for Y_1 memory			**SET/RESET for Y_5 memory**		
	L_1L_4	L_2L_3		$L_2L_3Y_4$	$L_2L_3Y_4'$
Y_6'	L_2L_4	L_2L_3		$L_2L_3Y_3$	$L_2L_3Y_4'$
Y_3	L_1L_4	L_2L_4		L_1L_3	$L_2L_3Y_4'$
SET/RESET for Y_2 memory				L_1L_4	$L_2L_3Y_4'$
	L_2L_3	$L_1L_4Y_6$		$L_2L_3Y_4$	$L_1L_3Y_2'$
	$L_2L_3Y_4'$	$L_1L_6Y_6$	Y_4	$L_2L_3Y_3$	$L_1L_3Y_2'$
Y_4, Y_5	$L_1L_4Y_6'$	$L_1L_4Y_6$		$L_1L_3Y_2$	$L_1L_3Y_2'$
Y_4	L_2L_4	$L_1L_4Y_6$	Y_4	$L_1L_3Y_1$	$L_1L_3Y_2'$
	$L_2L_4Y_5'$	$L_1L_4Y_6$		L_1L_4	$L_1L_3Y_2'$
SET/RESET for Y_3 memory				$L_2L_3Y_4$	L_1L_4
				$L_2L_3Y_3$	L_1L_4
Y_6	L_1L_3	L_2L_4	Y_4	$L_1L_3Y_2$	L_1L_4
Y_1	L_1L_4	L_2L_4		$L_1L_3Y_1$	L_1L_4
SET/RESET for Y_4 memory				$L_2L_3Y_4$	$L_1L_3Y_1'$
				$L_2L_3Y_3$	$L_1L_3Y_1'$
Y_5	$L_1L_3Y_2$	$L_1L_3Y_2'$	Y_4	$L_1L_3Y_1$	$L_1L_3Y_1'$
Y_5	L_1L_4	$L_1L_3Y_2'$	Y_2, Y_4	$L_1L_4Y_6'$	$L_1L_4Y_6$
	L_2L_4	$L_1L_3Y_2'$	Y_4	$L_1L_4Y_6'$	$L_1L_3Y_1'$
	$L_2L_4Y_5'$	$L_1L_3Y_2'$		$L_2L_3Y_3$	$L_2L_3Y_3'$
Y_5	$L_1L_3Y_2$	L_1L_4		$L_1L_3Y_1$	$L_2L_3Y_3'$
Y_5	$L_1L_3Y_1$	$L_1L_3Y_1'$		$L_1L_4Y_6'$	$L_2L_3Y_3'$
	L_2L_4	$L_1L_3Y_1'$	**SET/RESET for Y_6 memory**		
Y_2, Y_5	$L_1L_4Y_6'$	$L_1L_4Y_6$	Y_3	L_1L_3	L_2L_4
Y_5	$L_1L_4Y_6'$	$L_1L_3Y_1'$		L_1L_3	L_2L_3
Y_2	L_2L_4	$L_1L_4Y_6$	Y_1'	L_2L_3	L_2L_4

SEQUENTIAL LOGIC DESIGN—NONCLASSICAL

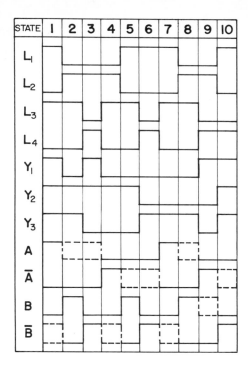

Figure 8-23 State diagram.

8-7 COMPUTER-AIDED SYNTHESIS

Although some of the procedures involved in the nonclassical methods are quite simple, very few have been computerized. The objective of such computer programs is quite evident—to avoid errors and to tolerate larger networks. In industries using fluid logic circuitry, the existence of such computer-aided design tools can be of major importance, especially for assisting the users in their varied applications.

In relation to nonclassical synthesis techniques, the authors have been associated with two computer programs:

1 The Change Signal Method.
2 The Total Signal Approach.

The computer program for the change signal method was developed by L. E. Bensch and is a computer-aided realization of Cole's method as discussed in Section 8-2. The program was designed in ANS Fortran IV and accepts the required sequence of operation as the input data to the program. The program returns the complete set of circuit equations to the user. For example, consider the two-cylinder system that is required to perform the sequence: A, B, $\bar{B}$, $\bar{A}$, B, A, $\bar{A}$, $\bar{B}$. The input to the program would be simply as follows:

card 1: 2 8 11 21 20 10 21 11 10 20

where the first two integers relate to the number of variables (cylinders) and states, respectively. The following integers represent the operation of the system, which are entered as a two-part integer *NM*, where *N* is the identification number of the variable and *M* is the EXTEND (1) or RETRACT (0) indicator for each operation. Thus, "11" indicates "cylinder 1 (or in this case, *A*) extending," while "20" relates to the retraction of cylinder 2 or *B*. The computer output for this example is shown in Table 8-11. The reader may note that no attempt was made to incorporate the prepared path implementation of the network.

Table 8-11 Computer Output of the Change Signal Computer Program

The output equations are:

$Z(1) = A'(2) \cdot Y(3) + A(2) \cdot Y'(4) \cdot W(2)$
$\bar{Z}(1) = A'(2) \cdot Y'(3) + A(1) \cdot Y'(2)$
$Z(2) = A(1) \cdot Y(2) \cdot W(1) + A'(1) \cdot Y(1)$
$\bar{Z}(2) = A(2) \cdot Y(4) + A'(1) \cdot Y'(1)$

The SET and RESET signals for the *Y* and *W* gates are:

$Y(1)$ SET $= A'(2) \cdot Y'(3)$
       RESET $= A(1) \cdot Y'(2)$
$Y(2)$ SET $= A'(2) \cdot Y(3)$
       RESET $= A(2) \cdot Y'(4) \cdot W(2)$
$Y(3)$ SET $= A'(1) \cdot Y'(1)$
       RESET $= A(2) \cdot Y(4)$
$Y(4)$ SET $= A(1) \cdot Y(2) \cdot W(1)$
       RESET $= A'(1) \cdot Y(1)$
$W(1)$ SET $= A'(2) \cdot Y(3)$
       RESET $= A(2) \cdot Y(4)$
$W(2)$ SET $= A'(1) \cdot Y(1)$
       RESET $= A(1) \cdot Y'(2)$

The modified operations table is:

ROW	EVENT	*	A		*	Y				*	W		*
			1	2		1	2	3	4		1	2	
1	11	0	0	0		1	1	0			1	0	
2	21	1	0	0		1	1	1			1	0	
3	20	1	1	0		1	0	1			0	0	
4	10	1	0	1		1	0	1			0	0	
5	21	0	0	1		1	0	0			0	1	
6	11	0	1	1		0	0	0			0	1	
7	10	1	1	0		0	0	0			0	0	
8	20	0	1	0		0	1	0			0	0	

SEQUENTIAL LOGIC DESIGN—NONCLASSICAL

Table 8-12   Input Data Set

Card	00000000011111111112...666666 12345678901234567890 345678
1	EXAMPLE PROBLEM
2	2 2 4 8
3	00 01 11 10
4	1001 2 0 3 00
5	11002 4 0 00
6	1 0 51003 00
7	0 21004 6 00
8	0 21005 3 01
9	1 0 71006 00
10	0 81007 6 00
11	11008 4 0 10

The second computer program utilizes the total signal concept for the resolution of logic problems and was developed by R. L. Woods. The reader should be alerted to the fact that this program is not a "direct mechanization" of any of the methods discussed in the preceding sections. The program was also written in ANS Fortran IV programming language and employs a concept similar to those utilized in Sections 8-3 and 8-4, except that the Primitive Flow Table is utilized as the logic specification format of the data input. Stable states in the table are indicated by the addition of "1000" to the state identification number, while unstable states are represented by the state number itself. For example, the problem discussed in Section 8-4 requires the data set as shown in Table 8-12. The first line of the data set indicates the number of inputs, outputs, and rows, respectively. The second line indicates the input states for each consecutive column of the table. Starting from the third line, the PFT is entered, row by row. "1007" refers to stable state 7, while all numbers less than 1000 represent the unstable states of the system. At the far right of the table, the output states are listed. The program produces the complete set of circuit equations as shown in Table 8-13.

PROBLEMS

1   Using the Change Signal Method, synthesize the problem represented by the preliminary operations table shown in Table P8-1.
   a   Write the circuit equations.
   b   Perform the necessary modifications to the circuit equations in order to obtain a prepared path network.
   c   Construct the network using three- or four-way spool valves.
   d   Construct a diagrammatic network using standard two-input logic symbols.
2   A three-cylinder system is to perform the following sequence:

   $P, \bar{P}, Q, \bar{Q}, R, \bar{R}, P, Q, \bar{Q}, R, \bar{P}, \bar{R}$

   a   Construct the Operations Table (use the INPUT-OUTPUT circuit as shown in Figure 8-1).

Table 8-13 Computer Output of the Total Signal Computer Program

Logic synthesis
for 2 inputs, 2 outputs

Original primitive flow table
for example problem

$A_1 A_2$ 00	01	11	10	Z_1	Z_2
(1)	2	0	3	0	0
1	(2)	4	0	0	0
1	0	5	(3)	0	0
0	2	(4)	6	0	0
0	2	(5)	3	0	1
1	0	7	(6)	0	0
0	8	(7)	6	0	0
1	(8)	4	0	1	0

Canonical flow table for
example problem

$A_1 A_2$ 00	01	11	10	Z_1	Z_2
(1)	3	0	2	0	0
1	0	4	(2)	0	0
1	(3)	5	0	0	0
0	3	(4)	2	0	1
0	3	(5)	6	0	0
1	0	7	(6)	0	0
0	8	(7)	6	0	0
1	(8)	5	0	1	0

(Passive memory assignment)

State signals:

(1) = 00
(3) = 01 * Y_1
(8) = 01 * Y_1'
(4) = 11 * Y_2 * Y_3
(5) = 11 * Y_2 * Y_3'
(7) = 11 * Y_2'
(2) = 10 * Y_4
(6) = 10 * Y_4'

Switching conditions:

Y_1 SET = 1 + 4 + 5
    RESET = 7
Y_2 SET = 2 + 3 + 8
    RESET = 6
Y_3 SET = 2
    RESET = 3 + 8
Y_4 SET = 1 + 4
    RESET = 5 + 7

Output signals:

$Z_1 = 8$
$Z_2 = 4$

SEQUENTIAL LOGIC DESIGN—NONCLASSICAL

Table P8-1

No.	Event	Changed signal	Synthesis
1	P	a'	
2	$\bar{P}$	a	
3	Q	a'	
4	$\bar{Q}$	b	
5	P	b'	
6	$\bar{P}$	a	
7	Q	a'	
8	P	b	
9	$\bar{Q}$	a	
10	$\bar{P}$	b'	

   b By the Change Signal Method, construct the nonprepared path network.
   c By the same method, construct the prepared path network.
3 Synthesize the network represented by the synthesis table of Table P8-3 using the Total Signal Method.

Table P8-3

No.	Outputs (P, Q)	Input state	Next state	Previous states	Synthesis
1	1 0	$a'b'$	2, 3	8, 9	
2	1 1	$a'b$	4, 6	1, 5	
3	0 0	ab'	5, 7	1, 6, 7	
4	0 1	$a'b'$	8, 10	2, 10	
5	1 1	ab	2, 9	3, 8	
6	0 0	ab	3	2, 9, 10	
7	1 1	$a'b'$	3	3	
8	0 0	$a'b$	1, 5	4	
9	0 1	ab'	1, 6	5	
10	1 1	ab'	4, 6	4	

   a Derive the minimal network equations.
   b Implement the network using two-input logic elements.
4 A network is represented by the LSC shown in Figure P8-4. Synthesize the network using the total signal method.
   a Construct the synthesis table.
   b Formulate the minimal network equations.
   c Implement the network using three- or four-way spool valves.
5 The control system represented by the operations table in Problem 1 is to be synthesized by the Total Signal Method.
   a Construct the synthesis table.
   b Synthesize the network and derive the minimal network equations.
6 By using the Total Signal Method, synthesize the control system of Problem 2.

Outputs		Inputs (a, b, c)					
P	Q	000	001	100	101	111	011
1	0	(1)	2	3			
1	1	1	(2)		4		
0	1	5		(3)	4		
0	0			3	(4)	6	
1	1	(5)	7	8			
1	0				4	(6)	9
1	0	5	(7)				9
1	1	1		(8)	10		
0	0		2				(9)
1	1		2	8	(10)	6	

Figure P8-4

7 A network is represented by the state matrix shown in Figure P8-7.

Output matrix Network matrix Input matrix

$$\begin{bmatrix} Z_1 \\ Z_1' \\ Z_2 \\ Z_2' \end{bmatrix} = \begin{array}{|c|c|c|c|} \hline 1(2,3) & 8(1,4) & 7(6,8) & 9(1,5) \\ \hline & 2(1,7) & 4(2,9) \atop 5(2,3) & 3(1,7) \atop 6(1,7) \\ \hline & 2(1,7) & 5(2,3) \atop 7(6,8) & 6(1,7) \atop 9(1,5) \\ \hline 1(2,3) & 8(1,4) & 4(2,9) & 3(1,7) \\ \hline \end{array} \quad \begin{array}{c} (a\ b) \\ \begin{bmatrix} 0 & 0 \\ 0 & 1 \\ 1 & 1 \\ 1 & 0 \end{bmatrix} \end{array}$$

Figure P8-7

   a Obtain the network equations.
   b Implement using two-input logic elements.
8 Obtain the state matrix representation of Problem 2 and derive the network equations.
9 Construct the state matrix representation of the LSC in Problem 4 and obtain the network equations.
10 The LSC shown in Figure P8-10 is to be synthesized using the transition diagram approach. Obtain the minimal solution, and implement the network using all available logic elements.
11 Use the transition table approach to synthesize the networks represented in Problems 1 through 4 (construct the LSC's of the problems).
12 Use the Total Signal Method to synthesize the control network represented in Problem 10.

SEQUENTIAL LOGIC DESIGN—NONCLASSICAL

Outputs			Inputs (a, b, c, d).						
P	Q	R	0101	0100	0110	0010	1010	1000	1001
0	1	0	(1)	2					
0	1	0	1	(2)	3				
1	1	0			(3)	4			
1	1	0			5	(4)	6		
0	0	1		7	(5)				
1	0	0					(6)	8	
0	0	1	1	(7)					
1	0	0					9	(8)	10
0	0	1					(9)	8	
1	1	0						11	(10)
1	1	0					12	(11)	
0	1	0				13	(12)		
0	1	0			14	(13)			
0	0	0		7	(14)				

Figure P8-10

13 Use the state diagram technique to synthesize a network having the synthesis table shown in Table P8-13.

Table P8-13

State	Feedback signal combination	Operation	Synthesis
1	$L_1 L_3$	P	
2	$L_2 L_3$	$\bar{P}$	
3	$L_1 L_3$	P	
4	$L_2 L_3$	Q	
5	$L_2 L_4$	$\bar{Q}$	
6	$L_2 L_3$	Q	
7	$L_2 L_4$	$\bar{P}$	
8	$L_1 L_4$	$\bar{Q}$	
9	$L_1 L_3$	P	
10	$L_2 L_3$	$\bar{P}$	

Obtain the minimal network representation.

14 Using the input-output circuit as presented in Figure 8-19, synthesize the control system presented in Problem 2.

DEFINITION

Feedback inputs Inputs that indicate the completion of a certain machine operation.

REFERENCES

Bensch, L. E., "A Computer Program for Synthesis of Regularly Activated Fluid Logic Sequential Circuits," Basic Fluid Power Research Conference, Computer-Aided Design Supplemental Report No. 70-3, Fluid Power Research Center, Oklahoma State University, Stillwater, Oklahoma, pp. 121–145, July 1970.

Cole, J. H., "Synthesis of Optimum Complex Fluid Logic Sequential Circuits," Ph.D. Dissertation, Oklahoma State University, 1968.

Cole, J. H., and Fitch, E. C., "Synthesis of Fluid Logic Control Circuits," Paper No. III-C4, 10th Joint Automatic Control Conference, Boulder, Colorado, Aug. 1969.

_____, "Synthesis of Fluid Logic Circuits with Combined Feedback Input Signals," American Society of Mechanical Engineers, Fluidics Conference, Atlanta, Georgia, Paper No. 70-Flcs-18, ASME, June 1970.

_____, "Synthesis of Fluid Logic Networks with Optional Input Signals," ASME, Fluidics Conference, Atlanta, Georgia, Paper No. 70-Flcs-19, ASME, June 1970.

Chen, P. I., and Lee, Y. H., "State Diagram Synthesis of Fluidic Feedback Circuits," ASME Winter Annual Meeting, Fluidics Committee, Detroit, Michigan, Paper No. 73-WA/Flcs-2, ASME, Nov. 1973.

Cheng, R. M. H., and Foster, K., "A Computer Aided Design Method Specially Applicable to Fluidic-Pneumatic Sequential Control Circuits," ASME Winter Annual Meeting, Fluidics Committee, New York, Paper No. 70-WA/Flcs-17, ASME, Nov. 1970.

Fitch, E. C., "Fluid Logic," Teaching Manual Published by The School of Mechanical and Aerospace Engineering, Oklahoma State University, Stillwater, Oklahoma, 1966.

Krieger, M. *Basic Switching Circuit Theory*. New York: The Macmillan Company, 1967.

Maroney, G. E., "A Synthesis Technique for Asynchronous Digital Control Networks," M. S. Report, Oklahoma State University, Stillwater, Oklahoma, 1969.

Maroney, G. E., and Fitch, E. C., "Stochastic Type Digital Fluid Control Systems Synthesis," Controls and System Conference, Chicago, Illinois, May 1970.

_____, "A Synthesis Technique For Fluid Logic Control Networks," Second Annual Fluid Logic Conference, Boston, Mass., April 1970.

_____, "Computer Aided Synthesis of Fluid Logic Systems," 26th National Conference on Fluid Power, Chicago, Illinois, Oct. 1970.

Surjaatmadja, J. B., "A Generalized Method For Synthesizing Optimal Fluid Logic Networks," Fluid Power Research Center, Report No. R73-FL-3, Oklahoma State University, Stillwater, Oklahoma, Oct. 1973.

Woods, R. L., "The State Matrix Method for the Synthesis of Digital Logic Systems," M. S. Thesis, Oklahoma State University, Stillwater, Oklahoma, 1970.

_____, "Users' Guide for the Time-Share Version of Logic Synthesis Program LOGSYN," Basic Fluid Power Research Conference, Computer Aided Design Supplemental Report No. 70-3, Fluid Power Research Center, Oklahoma State University, Stillwater, Oklahoma, pp. 81–119, July 1970.

Anonymous, "Air Logic Seminar Workbook," File No. 6215-174, Miller Fluid Power Corporation, Bensenville, Illinois, 1974.

Chapter 9

Network Analysis and Revision

9-1 ESOTERIC REALITY OF ANALYSIS

In many cases, a network is the result of intuitive reasoning and it exhibits no formal logic synthesis basis. Once such a network has been designed and is no longer associated with its creator, the job of modifying the system or troubleshooting can become a hectic experience. The resolution of a complex network requires a systematic analysis method which can reveal the logic implications and the inherent limitation of the system. Such a process can reveal the actual control capability of an existing network, identify redundant functions, and serve as the foundation for troubleshooting the operation of a machine.

A logic analysis is designed to separate and examine the individual elements of a system and determine whether their functional and integrated relationships are capable of satisfying the prescribed logic requirement. The main purpose of an analysis is to either verify the correctness of the logic system or to reveal the cause of integrated complications. To accomplish the purpose of an analysis requires a complete specification of the intended operation upon which a factual description of the existing network can be superimposed.

Once a flaw in the design has been discovered, it is obviously necessary to make revisions in the network. These revisions can be performed using two approaches:

1. Corrections upon the outputs (both primary and secondary) of the existing network.
2. Resynthesizing of the control network.

Corrections upon the outputs of the existing networks usually require additional elements and may not be a justifiable action when the required correction is severe. On the other hand, resynthesizing of the control system may be undesirable due to the difficulties that may arise in the synthesis, implementation and reanalyzing of the resulting network. Seldom can assurance be given that a resynthesized network is "better" than the original network. Nevertheless, resynthesizing is the only approach that should be considered when the network is to be mass-produced and the absolute minimal solution for cost reduction purposes is warranted. This chapter is concerned with techniques for analyzing networks and the effectiveness of error-correction schemes.

9-2 ANALYSIS OF COMBINATIONAL LOGIC NETWORKS

Analysis can be considered as the opposite of synthesis—a network exists and the task is to determine the operation of the machine. A comparison with the intended operation of the control network establishes whether the network possesses the necessary correctness of the factual operation. In combinational logic networks the desired assignment of the control network is recorded in truth table fashion. Therefore, the main objective of the analysis would be determining the truth function of the network under consideration.

One important step in an analysis is the development of the equations representing the implemented network. This step should not impose any difficulties to the reader, as he is by now quite familiar with implementing such network equations, especially when standard logic elements are utilized. The process is now reversed. An analysis starts where the inputs enter the networks. The outputs of each element having known inputs are properly labeled by their proper logic representation. This labeling process is performed until all outputs are identified. For example, consider the network shown in Figure 9-1. The elements in this figure are subsequently numbered in order to facilitate the discussion. For AND valve 1, the inputs are known; they are a and b. Therefore, the output of the AND valve can be determined, which is $a \cdot b$, and this is indicated at the output of element 1. Similarly, the output of the NOR element 3 is $a'c'$. Since (ab) and (cd) are the inputs to the OR element 5, the output of this element is $(ab + cd)$. After all the intermittent lines have been labeled properly, the output of the network can be written as follows:

$$Z = (ab + cd) \cdot (a'c' + bd') = abd' \tag{9-1}$$

NETWORK ANALYSIS AND REVISION

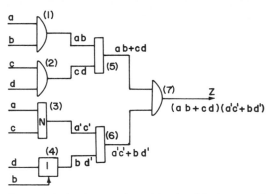

Figure 9-1 Implemented combinational logic network.

The truth table of the network can be observed in Table 9-1. The representation of the network should be written in its disjunctive form if the prescribed truth table has been constructed in a disjunctive form. This permits comparisons to be made. It should be noted that once the actual truth table has been constructed, the task of comparing the tables is quite trivial. The use of Karnaugh maps as a powerful tool for comparison purposes should be considered.

The above approach is valid even when networks having all NOR/NAND logic elements are encountered. However, in order to avoid the task of complementing and recomplementing each signal, it may prove advantageous to perform the task of complementation after the complete expression of the equation has been developed. The uncomplementation algorithm discussed in Chapter 2 may prove useful in

Table 9-1 Truth Table

a	b	c	d	Z
0	0	0	0	0
0	0	0	1	0
0	0	1	1	0
0	0	1	0	0
0	1	1	0	0
0	1	1	1	0
0	1	0	1	0
0	1	0	0	0
1	1	0	0	1
1	1	0	1	0
1	1	1	1	0
1	1	1	0	1
1	0	1	0	0
1	0	1	1	0
1	0	0	1	0
1	0	0	0	0

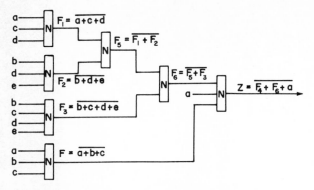

Figure 9-2 Implemented NOR-logic network.

performing this task. For example, for the circuit shown in Figure 9-2, the output (Z) is:

$$Z = \overline{\overline{a+c+d} + \overline{b+d+e} + \overline{b+c+d+e} + a} + \overline{a+b+c}$$
$$= ((a+c+d)(b+d+e) + b'c'd'e') \cdot a' \cdot (a+b+c)$$
$$= a'bc + a'bd + a'ce + a'cd \tag{9-2}$$

A more complex situation often occurs when the network contains nonconventional "logic" elements. For example, as mentioned in Chapter 2, a logic function can be satisfied using a composition of check valves. In such networks, the logic functions of a certain group of devices may not always be apparent. It is, therefore, necessary that the network be inspected at each point in the circuit in order to establish the logic expression for the particular point relative to the contributing variables. It should be noted that most check-valve logic exhibits a memory function; e.g., the function of the output is a function of its current state. However, the reader should realize that this "memory function" is a "powerless" memory, since it is established by a passive pressure which is held at a certain pressure (without power). Consequently, any leaks in the output line would disable the memory, and, therefore, caution should be exercised when such elements are incorporated in a network. Four basic functions of check valves used in performing logic are shown in Figure 9-3.

Oftentimes, the outputs of check valves are interconnected to each other. When such cases are encountered, it is necessary that these valves be considered simultaneously. For example, consider the two circuits shown in Figure 9-4, with their associated Karnaugh maps. Examination and simplification of the maps reveal that the first circuit (Figure 9-4(a)) is an AND element, while the second circuit (Figure 9-4(b)) is represented by the following representation:

$$Z = A \cdot B + A \cdot C + A \cdot Z \tag{9-3}$$

NETWORK ANALYSIS AND REVISION

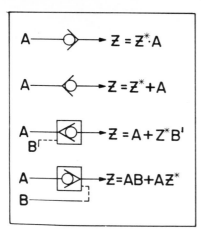

Figure 9-3 Logic representations of check valves.

It should be realized that a drainage of power can occur in both of the constructions of Figure 9-4 in some input states.

9-3 COMBINATIONAL LOGIC NETWORK REVISION

As mentioned earlier, the main objectives performing an analysis are to verify the operation of a network and to detect errors which could jeopardize the function of

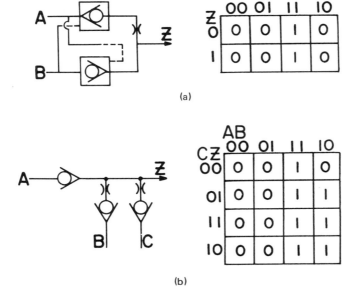

Figure 9-4 Check valve circuits and their Karnaugh map representation.

the machine. Once an error is detected, it is obvious that the objective should be extended toward the correction of the network. Resynthesizing may be the obvious solution; however, this approach should be avoided unless it proves to be the only rational answer to the problem. In many situations, it is economically and politically desirable to salvage an existing network. Such an approach is definitely advantageous when the network is "one-of-a-kind," as it avoids the time-consuming task of resynthesizing, implementing, and reanalyzing the network.

In the course of revising combinational logic networks, there are two basic types of errors that need consideration:

1 When the output is activated during undesired input states.
2 Where the output is OFF during a time that the input states and the system logic require the activation of the output.

The first situation may be illustrated by the Karnaugh map shown in Figure 9-5. The intended operation of the machine (as depicted by its truth table) is represented by the 1's, 0's and "don't cares" shown in the map. Assume that the actual operations of the implemented machine is represented in the map by the shaded area shown. From the map, it is apparent that the inclusion of ABCD = (1110) by the network would jeopardize the intended operation of the machine, if not corrected.

In order to avoid increasing the complexity of the network during error-correcting exercises, it is important for the reader to be aware of the possible cases which may exist. These cases are as follows:

1 The generation of the unintended "1" is apparent or recognizable and is no contribution toward the generation of the desired 1's.
2 The generation of the unintended 1 is not apparent.
3 The generation of the unintended 1 is apparent, but is involved with the generation of intended 1's.

Case 1 can be illustrated by the example depicted by the Karnaugh map in Figure 9-6. Again, the actual behavior of the network is represented by the shaded area shown. The network is represented by the following equation:

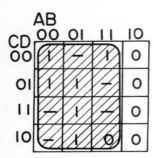

Figure 9-5 Karnaugh map showing an unintended "1."

NETWORK ANALYSIS AND REVISION

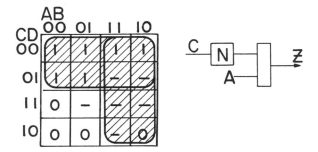

Figure 9-6 An undesired 1 generated by an unnecessary cube and its circuit.

$$Z = A + C' \tag{9-4}$$

From the map, it can be seen that the 1-state at $ABCD = (1010)$ is actually required to be 0; therefore network corrections must be performed. Further inspection reveals that the generation of this unwanted 1-state is made by the cube represented by A, and as all intended 1's covered by A are already covered by the only other cube, the network portion representing the cube A is redundant and therefore can be eliminated.

A different situation occurs when the generation of the unintended 1's is not apparent. This is the most general situation encountered in the analysis of logic networks. In such cases, the necessary modifications are not severe, since a simple "inhibiting" or "sharping" of the signals is all that is needed. This can be performed by applying the following rules:

1. Select the "unwanted 1's" as the 1's of the inhibiting function.
2. Select all 0's and "-'s" of the intended function as the "don't cares" of the inhibiting function.
3. Simplify the inhibiting function.
4. Implement the results and use the inhibiting output as the signal to "inhibit" or "sharp" the erroneous circuit output (Adds an INHIBITOR element).

In order to illustrate the above procedure, consider the situation shown in Figure 9-5. Taking the (1110) position as the 1 while all 0's and "-'s" are considered "don't cares," the inhibitor function is:

$$Z_1 = A \cdot C \tag{9-5}$$

By considering the network implemented previously as a "black-box," the network modification can be illustrated as shown in Figure 9-7. As the original network is represented by $A' + B$, the network resulting from the inhibiting process can be represented by the following functional description:

$$Z = (A' + B)(A' + C') = A' + BC' \tag{9-6}$$

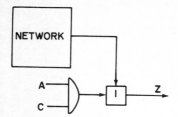

Figure 9-7 Network correction.

It can be noted that this representation satisfies the intended requirements of the machine.

Case 3 is a combination of the previously mentioned cases, and thus can be resolved by applying either of the earlier approaches. It should be recognized that one approach may be more convenient than the other, depending upon the nature of the problem. For example, in the problem illustrated by the Karnaugh map of Figure 9-8, two approaches may be taken: first, by eliminating the cube representation for *BD* and replacing it by cube $C'D$; or second, by inhibiting the network with the inhibiting cube *AC*. Both schemes can be observed in Figures 9-9 and 9-10 respectively.

If the output is OFF while the input states actually demands the activation of the output (error type 2), the solution is obvious, i.e., generate the output at the unaccessed input states. This can be performed most effectively by generating the "missing 1's" using the following rules:

1 Select the "missing 1's" as the 1's of the complementary expression.
2 Select the 1's and "-'s" as the "don't cares" of the expression.
3 Simplify the results and connect the output disjunctively to the erroneous network.

For example, consider the network represented by the Karnaugh map shown in Figure 9-11. The actual implemented network representation is again shown by the shaded area in the map. The complementary function can therefore be obtained and written as follows:

$$Z_C = C'D' + BCD \qquad (9\text{-}7)$$

The corrected network is illustrated in Figure 9-12.

9-4 SEQUENTIAL NETWORK ANALYSIS

Fundamentally, the process of synthesizing sequential networks can be divided into two distinct parts—the transformation of the sequential logic characteristics of the problem into several combinational-type logic problems and the actual synthesis of the combinational logic problems themselves. Since the process of analysis is in reality a true reversal of synthesis, it is reasonable that the same philosophy can be

NETWORK ANALYSIS AND REVISION

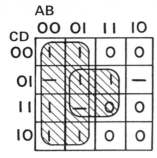

Figure 9-8 An inverted 1 generated by a necessary cube.

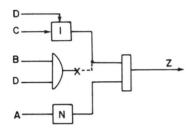

Figure 9-9 Network correction by replacement.

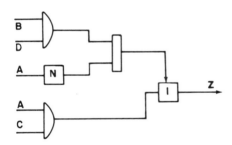

Figure 9-10 Network correction by inhibiting.

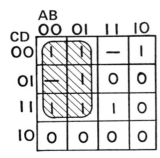

Figure 9-11 Karnaugh map showing "missing" entries.

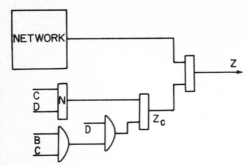

Figure 9-12 Network correction using complementary function.

applied for analyzing a sequential logic system. At this point, it is important to note that a sequential network is associated with the use of memory functions, whether or not conventional memory devices are incorporated. The reader is reminded of the fact that a memory function can be achieved by using not only basic logic elements but also unconventional memory devices such as check valves, delay elements, etc. In such situations, the problem of detecting memory functions in an existing circuit may prove to be difficult.

In general, a memory circuit is a circuit where its output is a function of its current state. Therefore, in circuits using conventional nonmemory-type logic devices, such circuits are characterized by feedback lines from their outputs (secondary outputs). These feedback lines are mandatory in order for the memory function to be satisfied (see Figure 9-13). On the other hand, although some feedback characteristics may be exhibited by a Flip-Flop circuit such as shown in Figure 9-14, these feedback lines are not an essential requirement for the circuit to be classified as a memory circuit.

A different situation exists when nonconventional logic devices are utilized in a circuit. For example, when check valves are employed, it may be assumed that they serve some type of memory function due to the fact that they are "one-way" flow elements. Yet, they may be providing only a combinational logic function, as for the example shown by the AND circuit of Figure 9-4.

The structure of the network equations is the same as for the case of combinational networks, with the exception of the memory functions involved in the system. As in network synthesis, the representation of the memory expressions are

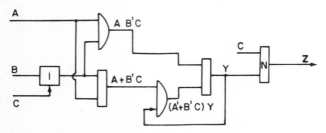

Figure 9-13 A typical feedback-type memory circuit.

NETWORK ANALYSIS AND REVISION

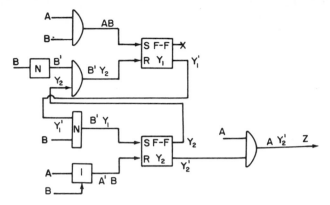

Figure 9-14 A typical sequential network.

treated separately from those of the outputs. Wherever memory devices are utilized in the circuit, it is beneficial to derive the SET and RESET equations of the memory devices as individual, combinational logic-type functions. By properly labeling each output of the memory elements, the task of determining the SET and RESET equations should pose little if any difficulties. For example, the network shown in Figure 9-14 has been labeled including the secondary variables. The network can hence be represented by the following set of network equations:

$$\text{SET}_{Y_1} = A \cdot B \qquad \text{RESET}_{Y_1} = B' \cdot Y_2 \qquad (9\text{-}8)$$
$$\text{SET}_{Y_2} = B' \cdot Y_1 \qquad \text{RESET}_{Y_2} = A' \cdot B \qquad (9\text{-}9)$$
$$Z = A \cdot Y'_2 \qquad (9\text{-}10)$$

As has been mentioned previously, a circuit which does not exhibit memory devices per se may still provide the function of a memory device. Feedback lines are definite indications of the presence of a memory function. Such lines should be labeled properly by memory Y-type designations. Similarly, unknown elements are best explored by first assuming them to be memory devices. This is accomplished by including the condition of the output as one of the influencing variables for the output itself. For example, by labeling the feedback line of Figure 9-13 by Y and labeling the other lines appropriately, the equations of the network are:

$$Y = AB'C + (A' + B'C) \cdot Y \qquad (9\text{-}11)$$
$$Z = C' \cdot Y' \qquad (9\text{-}12)$$

A more complex situation is often encountered when check valves are utilized in the circuit. The reader should be aware of the fact that these "powerless" memory devices are many times used as imperfect logic elements while completely ignoring their inherent memory capabilities. For example, the circuit of Figure 9-15 can be represented by the powerless memory element Y or by the imperfect

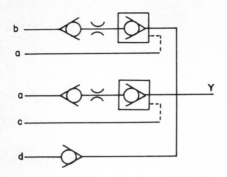

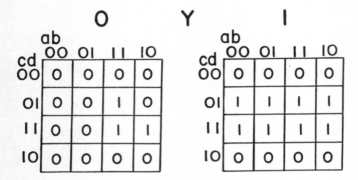

Figure 9-15 A typical check valve circuit and its Karnaugh map representation.

AND/OR element as represented by the following equations:

$$Y = (ab + ac) \cdot d + d \cdot y \tag{9-13}$$

or

$$Y = (ab + ac)d = (b + \dot{c}) a \cdot d \tag{9-14}$$

When this circuit is utilized as a memory variable, extreme care should be taken to insure that no flow is required for performing the logic (for example, by connecting the output to the control ports of downstream elements). On the other hand, when it is necessary to perform a combinational function in a sequential network, the circuit should be arranged such that input signal d is always deactivated prior to the deactivation of the other signals.

Once the primary and secondary outputs of the network have been determined, the network charts are developed in order to be able to facilitate the necessary assessments. The formation of the charts of the primary outputs is straightforward

NETWORK ANALYSIS AND REVISION 287

and is accomplished in the same manner as performed in the previous sections, as they are strictly combinational. The formation of the network logic charts is accomplished using the following steps:

1. Develop the secondary maps as follows: If the memories are only reflected by their algebraic representations (no actual memory elements were used) then:
   a. Form a chart having 2^N rows, where N is the number of such memories used in the system. The number of columns in the table is the number of input combinations that can be reached by the machine.
   b. Label each row and column by appropriate memory and input states.
   c. Project the memory equation in the chart by placing a 1 in each chart location where the corresponding states satisfies the memory representation.

   If the memories are represented by conventional SET and RESET expressions (memory elements are used), then perform the following steps:
   c. Project the SET and RESET equations in the chart by placing an S and/or R in each chart locations corresponding to the memory SET and/or RESET representation, respectively.
   d. Place a 1 in each location where only an S is found. Similarly, place a 0 in each location where only an R is found (for the particular memory element).
   e. Whenever the S and the R are both present, resolve the difference by consulting the truth table for the type of memory element involved.
   f. Whenever no S's and R's are present, the memory element is assumed to remain stable in that location; hence an entry that is equivalent to its current state (= row label) is entered.
2. Perform Step 1 for all memory elements.
3. Form the complete excitation map by "merging" all the above developed maps into one map.
4. In the resulting map, each entry having a binary value equivalent to its row designation is bracketed in order to indicate a stable state.
5. Obtain the Operational Flow Chart by:
   a. Assigning a decimal value to each bracketed binary number and retain the brackets.
   b. Assigning decimal values to each of the unbracketed entries by using numbers corresponding to the values of the stable-state numbers where the unstable conditions terminate.
6. Obtain the output map by reflecting the output equations on a map equivalent to those used for the secondaries.
7. Construct the Reduced Specification Chart by "expanding" the Operational Flow Chart. This is achieved by taking one stable state at a time and assigning them to separate rows. Each of the remaining states accessible to this particular stable state is positioned appropriately as an unstable state. On the left side of the chart, the corresponding outputs are listed.
8. Compare the reflected logic operation of the constructed RSC with the prescribed LSC. Comparisons should be performed one state at a time by observing whether or not the required outputs and next-state conditions are satisfied. The reader should note that state numbers assigned in the analysis and those of the logic specification (LSC) are most likely different since they were assigned in a random manner.

	ABC							
Y	000	001	011	010	110	111	101	100
0	(0)	(0)	(0)	(0)	(0)	(0)	1	(0)
1	(1)	(1)	(1)	(1)	0	0	(1)	0

Figure 9-16 Excitation map for Equation 9-11.

In order to illustrate the analysis procedure, consider the circuit shown in Figure 9-13. The circuit equations for this particular circuit have been presented earlier, see Equations 9-11 and 9-12. Since there is only one memory function, the excitation map is formulated as shown in Figure 9-16. Assuming that all input combinations are accessible, the chart is constructed using two rows and eight columns. The map is then inspected to identify the stable state locations of the map. Every 0 in the first-row locations ($Y = 0$) is designated as stable state, as are the 1's located in the second row. Numbering these states consecutively allows the OFC shown in Figure 9-17 to be constructed. Note that the unstable states are assigned appropriate next-state designations.

The output map is constructed almost independently from the excitation map. Its only connection to the excitation map is the format of the map itself—all row and column labels should be identical. Projecting the output equation on the map results in the output map shown in Figure 9-18. At this point, the operation of the implemented circuit has been determined. Comparisons with the desired operations depicted on the prescribed LSC can be made at this stage. However, to make comparative inspections more convenient, it may prove more effective to develop the RSC of the network representation. This is performed by taking one stable state at a time and selecting all entries of the associated row as unstable states. If there are restrictions regarding the input changes, such as where one input state cannot occur after another, these conditions can be appropriately reflected by eliminating "invalid" entries from the row.

For example, in the problem under consideration, if only single-input changes were permitted, the RSC would be constructed as shown in Figure 9-19. Note that unstable states 3, 5, 6, and 7 have been removed from row 1, as they are inaccessible from state 1. Similarly, unstable states 4, 5, 6, and 8 have been removed from the second row. The outputs of the system which are listed on the left side of the table were obtained by inspection of the output map.

A slightly different situation occurs when the excitation function is represented by SET and RESET equations. In this case, the actual "behavior" of the memory elements must be identified. In particular, it is important to know the predictable

	Input states (ABC)							
Y	000	001	011	010	110	111	101	100
0	(1)	(2)	(3)	(4)	(5)	(6)	7	(8)
1	(12)	(11)	(10)	(9)	5	6	(7)	8

Figure 9-17 Operational Flow Chart.

NETWORK ANALYSIS AND REVISION

```
      ABC
   Y  000 001 011 010 110 111 101 100
   0 | 1 | 0 | 0 | 1 | 1 | 0 | 0 | 1 |
   1 | 0 | 0 | 0 | 0 | 0 | 0 | 0 | 0 |
```

Figure 9-18 Output map.

condition of the element when it is exposed to different SET and RESET signal combinations. Generally, in fluid logic systems, extreme care should be taken when both SET and RESET signals exist during one or more machine states, since memory devices differ from one another regarding their output characteristics under such circumstances.

To illustrate the case where SET and RESET equations can be written, consider the network shown in Figure 9-14. The equations for the network are described by Equations 9-8, 9-9, and 9-10. The excitation map is constructed in two steps—first, by projecting the SET and RESET function, and second by interpreting the responses of the memories in relation to the SET and RESET conditions. The representation of the SET and RESET functions of Equations 9-8 and 9-9 in the "preliminary" excitation map can be observed in Figure 9-20. Note that subscripts (1) and (2) have been utilized for denoting memories Y_1 and Y_2, respectively. By inspection, it can be seen that no opposing memory signals exist throughout the map, and hence, memory excitation can be determined without requiring knowledge of the type of memory elements used.

For example, in the cube location (0000), the current memory state is (00); since no SET signals prevail, the memory elements are stable and should remain in that memory state (00). On the other hand, in cube location $ABY_1Y_2 = (1100)$, the SET signal for Y_1 is present, while no other signal prevails. Therefore, the next memory state for the Y_1 memory element should be (1), while the memory

Output (Z)	Input states (ABC)							
	000	001	011	010	110	111	101	100
1	(1)	2		4				8
0	1	(2)	3				7	
0		2	(3)	4		6		
1	1		3	(4)	5			
1				4	(5)	6		8
0			3		5	(6)	7	
0		11				6	(7)	8
1	1				5		7	(8)
0	12		10	(9)	5			
0		11	(10)	9		6		
0	12	(11)	10				7	
0	(12)	11		9				8

Figure 9-19 The Reduced Specification Chart.

	ab 00	01	11	10
Y_1Y_2 00	--,--;--,--	--,--;--,R_2	S_1,--;--,--	--,--,--,--
01	--,R_1;--,--	--,--;--,R_2	S_1,--;--,--	--,R_1;--,--
11	--,R_1;S_2,--	--,--;--,R_2	S_1,--;--,--	--,R_1;S_2,--
10	--,--;S_2,--	--,--;--,R_2	S_1,--;--,--	--,--;S_2,--

Figure 9-20 The development of the excitation map.

element Y_2 remains stable. Thus, a (10) is placed in this map location. The completed excitation map can be observed in Figure 9-21.

Once the excitation map has been developed, the OFC and the output map can be constructed in the same manner as in the previous example. The OFC for the example under consideration can be observed in Figure 9-22, while the output map is presented in Figure 9-23.

The task of comparing the analysis results with those of the actual prescribed logic specification is not as simple as in the case of combinational logic systems. This is due to the fact that in sequential systems, it is common for the same input state to occur in more than one machine state. However, the comparative part of the analysis assignment can be best performed using the following rules:

1 Select a column in the constructed Operational Flow Chart or Reduced Specification Chart which contains the least number of machine states.

Y_1Y_2 \ AB	00	01	11	10
00	(00)	(00)	10	(00)
01	(01)	00	11	(01)
11	01	10	(11)	01
10	11	(10)	10	11

Figure 9-21 The excitation map.

NETWORK ANALYSIS AND REVISION

		Input states (AB)			
Y_1	Y_2	00	01	11	10
0	0	(1)	(2)	4	(3)
0	1	(5)	2	6	(7)
1	1	5	8	(6)	7
1	0	5	(8)	(4)	7

Figure 9-22 The Operational Flow Chart.

2 Select a machine state in the corresponding column of the prescribed LSC and assume that this machine state is covered (or represented) by one of the states that is in the same column of the "new" OFC or RSC. If the associated column of the LSC is empty, select another column of the OFC or RSC.
3 Starting with the states selected in Step 2, try to follow the logic specification of the problem within the OFC or RSC. Note that it is possible for the RSC or OFC to have more or less machine states than the LSC, and that one machine state of the OFC can cover more than one machine state of the LSC and vice versa.
4 If no similarities exist, try to compare the state under discussion in the LSC with another state of the RSC, if any (in the same column).
5 Perform Steps 3 and 4 until a match is obtained.
6 If the specification of the problem matches the operation of the network, renumber the states of the OFC accordingly.
7 Failure to establish a match indicates the network being analyzed is incorrect.

To illustrate the rules for comparing the network logic with a specification consider the LSC shown in Figure 9-24 and its implemented circuit shown in Figure 9-25. By applying the rules discussed earlier in this section, the Operational Flow Chart and the output map can be formed as shown in Figure 9-26. Comparing the OFC with the LSC reveals that the number of machine states is not the same. Selecting the column that contains the least number of states, in this case the second column, trial comparisons can be performed.

In the LSC (Figure 9-24), this column contains two states, which are 2 and 4. Arbitrarily selecting state 2 for conducting the comparison, it can be observed that

$y_1 y_2$ \ AB	00	01	11	10
00	0	0	1	1
01	0	0	0	0
11	0	0	0	0
10	0	0	1	1

Figure 9-23 The output map.

INTRODUCTION TO FLUID LOGIC

Outputs		Input states (a, b)			
Z_1	Z_2	00	01	11	10
0	0	(1)	2	6	3
0	1	5	(2)	10	8
—	1	1	—	10	(3)
0	1	1	(4)	9	3
0	0	(5)	4	10	7
1	1	—	4	(6)	3
0	1	5	—	6	(7)
1	1	1	—	9	(8)
1	0	—	2	(9)	3
1	—	—	2	(10)	7

Figure 9-24 The Logic Specification Chart.

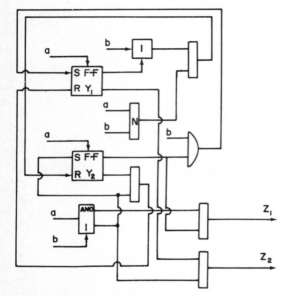

Figure 9-25 The implemented circuit.

		Input states (a, b)				Outputs (Z_1, Z_2)			
Y_1	Y_2	00	01	11	10	00	01	11	10
0	0	(1)	4	(2)	(3)	00	01	11	01
0	1	1	(4)	6	(5)	00	01	11	11
1	1	7	4	(6)	3	00	01	10	10
1	0	(7)	4	2	3	00	01	10	00

Figure 9-26 The Operational Flow Chart and Output Map.

NETWORK ANALYSIS AND REVISION

the outputs of the LSC and the OFC, as revealed by its output map at this particular state (state 4), match each other. The "next states" for the prescribed LSC state 2 are states 5, 10, and 8, which are represented in the OFC by states 1, 6, and 5, respectively. Comparing the outputs of the "next states" reveals that the output conditions are satisfied. At this point, it has been assumed that state 5 of the LSC is represented by the OFC state 1; further examination reveals that the "next states" in the LSC are states 4, 10, and 7, which are represented in the OFC as states 4, 2, and 3, respectively. Again, it can be shown that the outputs as defined by the LSC and the ones represented by the OFC do not contradict each other. However, it can be noticed that state 4 of the OFC has been used to represent state 4 of the LSC and this state has states 1, 9, and 3 as its "next state" requirements. In order to prove the validity of the OFC, it is important to show that these "next state" requirements can be represented by the OFC states 1, 6, and 5, respectively. The procedure is carried out until all loops in the LSC have been considered. It can be verified that the OFC shown in Figure 9-26 together with its related output map satisfies the required logic prescribed by the LSC of Figure 9-24. This is true even though the network contains a redundant OFC state 7 which is not indicated by the LSC.

9-5 SEQUENTIAL NETWORK REVISION

Once the errors in the machine logic have been detected, attention must be given to the modification of the network. As stated earlier, resynthesis of the network should be considered the last resort as the action for achieving the intended machine operation. The two basic reasons making network revisions necessary are:

1 Incorrect synthesis or implementation of the network.
2 Modifications required in altering the original network such as adding a new cycle.

Fundamentally, changes in sequential networks can be accomplished directly in the excitation and output maps, since such changes require the modification of the secondary and primary outputs of the system. These modifications can be performed by means of the approaches discussed in Section 9-3.

Since the synthesis of sequential networks satisfies the operational logic by formulating various combinational logic networks, there are many possibilities for errors or differences to occur. These possibilities can be conveniently classified into two major categories; i.e.:

1 Differences in the primary outputs.
2 Differences in the secondary outputs.

The first classification does not influence the cycle of the machine, and the problem in reality is "strictly combinational"; hence, the approaches discussed in Section 9-3 can be successfully applied. The second category requires modification of the machine cycle; therefore, extreme care must be exercised in order to obtain a correct, error-free and low-cost modified network.

When a modification in the sequence of the machine is required, the necessary changes should be made within the excitation map itself. Because of the large variety of problems that may exist in a sequential system, no set of definite rules can be formulated for the modification of the excitation variables. However, during the analysis, when the necessary modifications are being established, the following points should be considered in order to achieve a minimal network:

1. Always attempt to perform the modifications in the existing excitation map. This is accomplished by either:
   a. "Shifting" a stable state upward or downward in the excitation map, thus changing the secondary state of the particular machine state.
   b. Using the "unused" map locations. Unused map locations are those not employed in accomplishing logic of the machine.
2. Utilize new memory variables in order to create new and essential empty map locations.

It is obvious that the second approach must be avoided if possible, as new memory variables would require additional elements to implement the memories and the generation of their associated SET and RESET signals. In order to illustrate the procedure that is involved in the task of modifying a network, consider the network represented by the following equations:

$$A_1 = a'y_1' + a'b' \tag{9-15}$$
$$A_2 = b + ay_2' \tag{9-16}$$
$$B_1 = ay_2 \tag{9-17}$$
$$B_2 = b \tag{9-18}$$
$$\text{SET}_{Y_1} = ay_2 \qquad \text{RESET}_{Y_1} = ay_2' \tag{9-19}$$
$$\text{SET}_{Y_2} = a'y_1' \qquad \text{RESET}_{Y_2} = a'b'y_1 \tag{9-20}$$

The input-output circuit is as shown in Figure 9-27, where the input variables, a and a', are used to represent the extension and retraction of cylinder A, respectively. Assume the network was designed originally to satisfy the operation of a

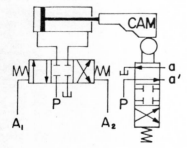

Figure 9-27 Input-output circuit.

NETWORK ANALYSIS AND REVISION

two-cylinder system having the following sequence:

$A, B, \overline{AB}, A, \overline{A} \ldots \ldots \ldots$

Note that the third operation is the simultaneous retraction of both cylinders.

The excitation and output maps can be constructed as shown in Figure 9-28. By examining the output map, the anticipated operations at each map location can be deduced and are listed within the map. Tracking the circuit operations through the map allows the implemented machine operations to be determined. For example, if it is assumed that the machine starts at map location $abY_1Y_2 = (0001)$, the anticipated output requires the extension of cylinder A and, therefore, brings the system to the stable location at location (1011). At this location, the output indicates that the operation is the extension of cylinder B. Examining the map further, it can be verified that the operation of the implemented machine is:

$A, B, \overline{A}, \overline{B}, A, \overline{A}, \ldots \ldots$

The state numbers are shown in Figure 9-28 in order to illustrate the operation in the map.

In order for the modified version of the circuit to satisfy the intended operation, it can be seen that one discrepancy exists and that is the condition of the outputs at state 3. At this state, the outputs should be such that the retraction of cylinder B is also initiated. This is performed by replacing the outputs of this particular location by $A_1 A_2 B_1 B_2 = (0101)$. However, the simultaneous retraction of both cylinders creates a "race" between the two cylinders—which could be critical if the retraction of cylinder B is completed before cylinder A. Under this circumstance, the system would find itself at state 2, requiring the extension of cylinder B, which definitely jeopardizes the intended operation of the machine.

Excitation ($Y_1 Y_2$)

(a)

$Y_1 Y_2$ \ ab	00	01	11	10
00	1 01	01	(00)	(6) (00)
01	(1) (01)	(01)	11	2 11
11	5 10	(4) (11)	(3) (11)	(2) (11)
10	(5) (10)	(10)	00	6 00

Outputs ($A_1 A_2 B_1 B_2$)

(b)

$Y_1 Y_2$ \ ab	00	01	11	10
00	1 1000 A	1101 $\overline{B}$	0101 $\overline{A}, \overline{B}$	(6) 0100 $\overline{A}$
01	(1) 1000 A	1101 $\overline{B}$	0111 $\overline{A}$	2 0010 B
11	5 1000 A	(4) 0101 $\overline{A}, \overline{B}$	(3) 0111 $\overline{A}$	(2) 0010 B
10	(5) 1000 A	0101 $\overline{A}, \overline{B}$	0101 $\overline{A}, \overline{B}$	6 0100 $\overline{A}$

Figure 9-28 (a) Excitation map; (b) output map.

296 INTRODUCTION TO FLUID LOGIC

$Y_1 Y_2$ \ ab	00	01	11	10
00	1 01	01	(00)	(6) (00)
01	(1) (01)	(01)	3 11	(2) (01)
11	5 10	(4) (11)	(3) (11)	(7) (11)
10	(5) (10)	(10)	00	6 00

Excitation $(Y_1 Y_2)$
(a)

$Y_1 Y_2$ \ ab	00	01	11	10
00	1 1000 A	1101 $\bar{B}$	0101 $\bar{A}, \bar{B}$	(6) 0100 $\bar{A}$
01	(1) 1000 A	1101 $\bar{B}$	3 01-- $\bar{A}(\bar{B})$	(2) 0100 $\bar{A}$
11	5 1000 A	(4) 0101 $\bar{A}, \bar{B}$	(3) 0101 $\bar{A}, \bar{B}$	(7) 010- $\bar{A}, \bar{B}$
10	(5) 1000 A	0101 $\bar{A}, \bar{B}$	0101 $\bar{A}, \bar{B}$	6 0100 $\bar{A}$

Outputs $(A_1 A_2 B_1 B_2)$
(b)

Figure 9-29 (a) Modified excitation map; (b) modified output map.

One way of satisfying the desired operation is to insure that map location (1011) have outputs that require the further retraction of cylinder A. This can be accomplished by moving state 2 somewhere else in the map. The reader should agree that the best alternative for placing state 2 is in location (1001), which requires that an unstable state 3 be placed in location (1101). The modification made in the maps are shown in Figure 9-29 by the boldface letters. The required changes in the network would be as follows:

$$SET_{Y_1} = (a \cdot y_2) \cdot b'' = (a \cdot y_2) \cdot b \qquad (9\text{-}21)$$
$$A_2 = A_{2(OLD)} + ay_1 \qquad (9\text{-}22)$$
$$B_1 = B_{1(OLD)} \cdot b' \qquad (9\text{-}23)$$

These network modifications can be observed in Figure 9-30, where the dashed lines indicate the network prior to the modifications.

Many times, it may become necessary for additional memory elements to be incorporated in the circuit. For example, assume that the above network is to be altered so that the following operation of the network occurs:

$$A, B, \overline{AB}, A, \bar{A}, A, \bar{A}, \ldots \ldots$$

It can be verified that it is impossible to change the excitation maps without incorporating new memory variables. Using one new memory, Y_3, results in the modified excitation and output maps shown in Figure 9-31. The necessary modifications in the network equations are:

$$SET_{Y_2} = ay_3 \qquad (9\text{-}24)$$
$$B_1 = B_{1(OLD)} \cdot b' \cdot y_3' \qquad (9\text{-}25)$$

NETWORK ANALYSIS AND REVISION

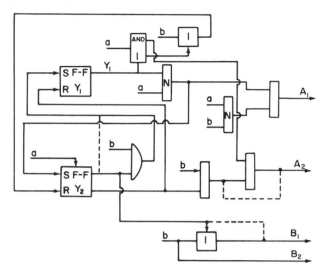

Figure 9-30 The network.

$Y_3Y_1Y_2$	ab 00	01	11	10
000	8 100	001	(000)	(6) (000)
001	(1) (001)	(001)	3 011	(2) (001)
011	5 010	(4) (011)	(3) (011)	(7) (011)
010	(5) (010)	(010)	000	6 000
110	---	---	---	---
111	---	---	---	---
101	1 001	---	---	(9) (101)
100	(8) (100)	---	---	9 101

(a)

$Y_3Y_1Y_2$	ab 00	01	11	10
000	8 1000 A	1101 $\bar{B}$	0101 $\bar{A}, \bar{B}$	(6) 0100 $\bar{A}$
001	(1) 1000 A	1101 $\bar{B}$	3 01-- $\bar{A},(\bar{B})$	(2) 0010 B
011	5 1000 A	(4) 0101 $\bar{A}, \bar{B}$	(3) 0101 $\bar{A}, \bar{B}$	(7) 010- $\bar{A},(\bar{B})$
010	(5) 1000 A	0101 $\bar{A}, \bar{B}$	0101 $\bar{A}, \bar{B}$	6 0100 $\bar{A}$
110	---	---	---	---
111	---	---	---	---
101	1 1000 A	---	---	(9) 0100 $\bar{A}$
100	(8) 1000 A	---	---	9 0100 $\bar{A}$

(b)

Figure 9-31 (a) Modified excitation map; (b) modified output map.

$$A_2 = A_{2(\text{OLD})} + ay_1 + ay_3 \tag{9-26}$$
$$\text{SET}_{Y_3} = a'y_1'y_2' \qquad \text{RESET}_{Y_3} = a'y_2 \tag{9-27}$$

9-6 TABULAR METHOD FOR NETWORK ANALYSIS

Quite often, a sequential network possesses a large number of memory elements, although the number of the desired machine states is quite limited. Such cases exist when the network was originally synthesized using approaches that did not stress the minimization of memory variables. Projecting such networks in the excitation maps as characterized by the classical synthesis would require an enormous map, due to the proliferation of the number of memory states with respect to the number of memory variables. Usually, unnecessary troubles arise due to the prevalence of a large number of unaccountable machine states.

Under such circumstances, the use of tabular methods involving the Operations or the Synthesis tables seems to be quite appropriate. Such tables do not require the individual representation of all memory states; therefore, the number of rows representing the machine is limited to the number of states prescribed for the machine. The synthesis table permits the recording of stochastic networks; therefore, its format is condoned and is referred to as the "analysis table."

As previously discussed, an analysis is initiated by the development of network equations. Memory functions should be identified, and their describing equations derived either in feedback form (i.e., the memory variable represented as a function of itself) or in a SET and RESET form (where memory elements are used). Once circuit equations are derived, the process of analysis can proceed using the following steps:

1. Formulate the preliminary analysis table of the *intended* operation. The synthesis table is constructed by listing the state numbers, intended outputs, input states, and the intended next states in subsequent columns.
2. Using the derived equations of the outputs, determine the conditions of the memory elements at specific machine states. Record these conditions in the columns assigned to the particular memory elements.
3. Inspecting each memory equation, determine whether or not the SET (or the activation) or the RESET (deactivation) signals are solely dependent on the inputs. Under such circumstances, the condition of the memory element at the known SET and RESET state can be determined. If the memory is represented by SET and RESET expressions, extreme care should be exercised, due to the possibility of having conflicting SET and RESET signals.
4. From each memory equation dependent upon the input states and known memory states, determine the condition at the known SET and RESET states.
5. Determine whether or not the results obtained in Step 3 conflict with the results of Step 2. Any conflicting condition indicates an erroneous operation of the system. In either case, the analysis proceeds by inserting the memory conditions obtained in Step 3 in the appropriate columns, thus subordinating those obtained in Step 2.
6. Starting with the machine state having the most memory conditions, perform the

simulation of the memory states throughout the machine cycle. Memory functions having **SET** and **RESET** features remain in the same condition unless a SET or RESET signal requires it to be otherwise. The conditions of memory functions constructed by feedback means are determined by observing their excitation. The determination of the memory conditions is initially performed relative to the past state of the memories. When switching occurs, it is necessary to reevaluate the conditions of all memory elements until a stable condition is obtained. At this point, all memory conditions which depend upon unknown conditions of certain memories should remain unknown. All memory states obtained in this step should dominate all memory conditions obtained previously.

7   Perform Step **6** at each point throughout the complete cycle of the machine.
8   If some memory conditions remain unknown, again select the machine state having the most known memory conditions; assume values for the unknown memories and perform the simulation as indicated in Step **6**.
9   After returning to the initial condition of the simulation, if the state of the memories is different than the state assumed at the beginning of the simulation, ignore the previously assumed state and continue the simulation for at least one more cycle.
10 If, after performing the simulation through a few complete machine cycles, the memory states of some machine states continue to change, or if the memory state of a machine state varies depending upon the previous state, then the particular "unstable" memory condition is assigned a "-" or "don't care."
11 Using the derived memory excitation and input states, formulate the implemented network for each machine state and compare them with the prescribed outputs. Only the prescribed "don't cares" can have different outputs.
12 The network analyzed is termed correct if all the outputs satisfy the condition states in Step **9**.

In order to illustrate the procedure, consider the two-cylinder problem discussed in the previous section. The intended cycle of the operation is $A, B, \overline{AB}, A, \overline{A}$.... The network equations have been presented in the previous section and are again presented here for the convenience of the reader:

$$A_1 = a'y'_1 + a'b' \tag{9-15}$$
$$A_2 = b + ay'_2 \tag{9-16}$$
$$B_1 = ay_2 \tag{9-17}$$
$$B_2 = b \tag{9-18}$$
$$\text{SET}_{Y_1} = ay_2 \qquad \text{RESET}_{Y_1} = ay'_2 \tag{9-19}$$
$$\text{SET}_{Y_2} = a'y'_1 \qquad \text{RESET}_{Y_2} = a'b'y_1 \tag{9-20}$$

The preliminary analysis table representing the logic of the problem can be observed in Table 9-2. Note that the simultaneous retraction of cylinders A and B at state 3 creates a race between the two cylinders. Therefore, to avoid erroneous operations, every possibility that may result from the race must be considered.

The analysis process is initiated by ascertaining the conditions of the memory

Table 9-2 The Preliminary Analysis Table

State	Desired output				Input		Next state	Analysis
	A_1	A_2	B_1	B_2	a	b		
1	1	0	0	—	0	0	2	
2	—	0	1	0	1	0	3	
3	0	1	0	1	1	1	4, 7	
4	0	—	0	1	0	1	5	
5	1	0	0	—	0	0	6	
6	0	1	0	—	1	0	1	
7	0	1	0	—	1	0	5	

elements using the output equations. This is performed by observing the output equations that are dependent upon one or more memory states and each condition of the memories that are absolutely necessary for the activation or deactivation of the particular output. The resulting condition is attributed to the condition of the memory at the associated states. For example, from the output equations, it can be noted that the equation of A_1 is represented by $a'y'_1 + a'b'$, while the prescribed logic of the network requires that A_1 be ON at both states 1 and 5. Therefore, at this point, it is already apparent that a redundancy exists; however, as the objective at this point is to analyze the network, the term $a'y'_1$ should be considered as a contributing signal to A_1. Observing Table 9-2 further reveals that output A_1 should be OFF at state 4, which could be activated if the memory element Y_1 is in the RESET condition. Therefore, Y_1 is required to be ON at state 4; and this information is recorded in the Y_1 column of the analysis table (see Table 9-3, shown by the underlined entry in the Y_1 column).

It can also be noticed that the equation for A_2 is dependent upon $b + ay'_2$, while the prescribed system requires the activation of output A_2 at states 3, 6, and 7. The activation of A_2 at state 3 is performed by the first term, b; therefore, the activation of the term ay'_2 is not compulsory at this state. However, during states 6 and 7, b is not activated, and hence, this leaves no alternative other than for ay'_2 to

Table 9-3 The Analysis Table

State	Desired output				Input		Next state	Memories		Actual output			
	A_1	A_2	B_1	B_2	a	b		Y_1	Y_2	A_1	A_2	B_1	B_2
1	1	0	0	—	0	0	2	0	1	1	0	0	0
2	—	0	1	0	1	0	3	1	1	0	0	1	0
3	0	1	0	1	1	1	4, 7	1	01	0	1	1	1
4	0	—	0	1	0	1	5	1	1	0	1	0	1
5	1	0	0	—	0	0	6	1	0	1	0	0	0
6	0	1	0	—	1	0	1	0	0	0	1	0	0
7	0	1	0	—	1	0	5	1	01	0	0	1	0

NETWORK ANALYSIS AND REVISION

be activated at this moment. As signal a is activated during these states the activation of signal ay'_2 can be satisfied by letting memory Y_2 be reset during these two states. Similarly, the output $B_1 = ay_2$ requires that memory Y_2 be activated at state 2, and that it be deactivated during states 3, 6, and 7, due to the fact that signal a is activated during these machine states (see underlined entries in Table 9-3).

As there are no memory SET or RESET signals that are independent of the states of the existing memories, the next step is to perform the simulation of the memory states throughout the machine cycle. For machines which possess stochastic characteristics, extreme care should be taken in order that every possible "path" or switching order is considered. From the table, it should be noted one memory condition has been predicted in states 2, 3, 4, 6, and 7; selecting state 2 randomly as the initial point of the simulation, the following points are apparent:

1. Having Y_2 activated at state 2, Y_1 must be activated due to the presence of signal ay_2 and the absence of its reset signal (ay'_2).
2. The next state of state 2 is state 3. No change of memory conditions should take place if no SET or RESET signals are present. Observing the table, no RESET signal is present in state 3.
3. From this, the memory state 3 is $Y_1Y_2 = (11)$, which overrides the existing 0-condition for Y_2 (indicating the existence of an error in the output).
4. The next possible states of state 3 are states 4 and 7. Referring again to state 4, no RESET signal is present, and, therefore, the memories remain stable at 11.
5. Similarly, at state 7, the memory state remains at 11, thus overriding the existing 0-condition of memory Y_2.
6. The next state for states 4 and 7 is state 5. At this state, memory element Y_2 resets due to the presence of the y_1 signal as well as the deactivation of both inputs.
7. At state 6, memory Y_1 resets due to the prevalence of the excitation signal ay'_2.
8. State 6 is followed by state 1, where Y_2 is activated due to signal $a'y'_1$.
9. At this point it is important to check whether the initial point (state 2) has the same memory state as determined at the beginning of the simulation. Any results different than the previously determined state will require that the simulation be continued for at least one complete cycle. In this example, Y_1 sets at state 2; it therefore satisfies the required initial condition and the simulation is terminated.

The completed excitation pattern of the memory elements can be observed in Table 9-3. Having determined all the memory states at each machine state, the actual expected outputs of the machine can be derived and listed in the analysis table. Observing the results, it is apparent that the actual outputs do not satisfy the required logic of the control system.

Having detected an erroneous operation of the machine, it is now important that the circuit be corrected. Avoiding the task of resynthesizing, the correction can be corrected by performing the following steps:

1. Observe the excitation pattern in the analysis table and determine the machine states that are not uniquely represented by the input and memory states.

2. Assign new memory variables to the nonunique input-memory states and determine the SET and RESET equations of these memories in the same manner as performed in the Total Signal Synthesis method discussed in the previous chapter.
3. Determine the necessary modifications that should be performed to correct the situation.

Evaluating the above steps, it is apparent that new memory elements are employed. This is due to the fact that the analysis table does not reveal the "unused" machine states that the excitation map would, and an attempt to derive such "unused" states in systems having many input and secondary variables may prove quite difficult.

In order to demonstrate the procedure for revising the network, consider the analysis table of the previously discussed problem (Table 9-3). The input and excitation patterns of the table reveal that states 2 and 7 are nonunique with respect to each other. Therefore, a new memory variable is introduced, together with its SET and RESET signals as shown in Table 9-4. The SET and RESET signals are derived as in the previous chapter, and are as follows:

$$\text{SET}_{Y_3} = b \qquad \text{RESET}_{Y_3} = a' \qquad (9\text{-}28)$$

The necessary corrections to the outputs can hence be accomplished as in the previous section, and they are:

$$A_2 = A_{2(\text{OLD})} + ay_3 \qquad (9\text{-}29)$$
$$B_1 = B_{1(\text{OLD})} \cdot b' \cdot y'_3 \qquad (9\text{-}30)$$

9-7 TROUBLESHOOTING

Even though a network has been properly synthesized and implemented, there are situations which can cause a network to perform undesired operations. These conditions can occur due to faulty elements or to system-imposed conditions such

Table 9-4 The Corrected Analysis Table

State	Desired output				Input		Next state	Memories			Actual output			
	A_1	A_2	B_1	B_2	a	b		Y_1	Y_2	Y_3	A_1	A_2	B_1	B_2
1	1	0	0	—	0	0	2	0	1	R	1	0	0	0
2	—	0	1	0	1	0	3	1	1	0	0	0	1	0
3	0	1	0	1	1	1	4, 7	1	1	S	0	1	1	1
4	0	—	0	1	0	1	5	1	1	—	0	1	0	1
5	1	0	0	—	0	0	6	1	0	—	1	0	0	0
6	0	1	0	—	1	0	1	0	0	—	0	1	0	0
7	0	1	0	—	1	0	5	1	1	1	0	0	1	0

NETWORK ANALYSIS AND REVISION

as clogged lines, leaks, etc. Performing mathematical analysis in such instances may prove to be a fruitless effort, whereas a slightly different type of analysis—an analysis concerned more with the actual behavior of the control network—may reveal the most effective action to pursue.

The task of troubleshooting includes an in-depth analysis of the actual machine behavior, establishing the cause of malfunction and correcting the faults in the circuit. A means must be available for monitoring the network operation at each point in the circuit. A few suggestions are offered below for detecting the faults in a control network:

1 If the network is a combinational-type logic network, perform the troubleshooting by the following approach:
   a Derive the mathematical expression of each element output that contributes toward the generation of the erroneous output (label each output line).
   b Simulate the network operation by tracking through the machine states and monitoring the second-level outputs of the networks.
   c Compare the results of the second-level monitoring with the required logic and determine the erroneous second-level output.
   d Monitor the third level, but only those third-level outputs that contribute to the generation of the erroneous second-level output.
   e This "homing in" procedure is performed at other levels in the same manner until the malfunction or error is detected. An error is found whenever all N-th level outputs are performing the intended operation, while its related $(N - 1)$th output is not.
2 For a sequential network, simulations should be performed in two parts: the simulation of the outputs with respect to the inputs and the secondary outputs; finally, the simulation of the secondary outputs themselves. If memory elements are utilized, perform the simulation of the SET and RESET signals individually, followed by the analysis of each memory element regarding their SET and RESET signals.
3 If no error is found in the sequential network (using Step 2), then perform the Classical Analysis in order to obtain the OFC of the control network. A thorough investigation should then be performed of this chart in order to discover the existence of essential hazards, d-trios, and all other N-th order hazards. Further adjustments of the delays existing in the feedback paths may then be necessary.

It should be realized that the above steps are outlined in a general manner, but should serve as effective guidelines for the troubleshooting of most combinational and sequential networks. The use of special devices such as one-shots, timers, and other accessories are extremely useful and should be considered as problem solvers.

9-8 COMPUTER-AIDED ANALYSIS

Although many scientists have devoted much time to computerize various synthesis methods, a surprising few have explored computer-aided methods dedicated to the analysis of fluid logic networks. Especially in relation to correcting erroneous logic networks, the authors have not encountered any program that is capable of performing this task mechanically. This void in the art of fluid logic is most obviously created by the large variety of errors that may occur in the synthesis and implementation process.

One method which was developed to analyze sequential fluid logic systems is credited to R. L. Woods. The Woods program was written in Fortran IV programming language. It requires that the network equations be derived manually and used to form the reduced flow table of the analyzed network. The program requires the following input data set:

 Card 1 : Identification card
    2 : The number of inputs, outputs, and memory elements.
    3 : The initial state of all input and memory variables.
    4 : Until card ($2^N + 3$), where N is the number of inputs: The number of memory elements that are associated with each input state.
 ($2^N + 4$) .... : The equations of the network, starting with the SET and RESET functions of each memory element. The equations are constructed as a function of ITS(i) (input states no. i) the inputs (X(1), X(2) .....) and the memories (Y(1), Y(2) .....). The recognition of the input states (ITS) in a logic equation may substantially reduce the computer and programming time. The i-th ITS is merely the i-th input state of the sequence when the inputs are coded in a Gray Code fashion. (($X(N)$, $X(N\text{-}1)$, ... $X(2)$, $X(1)$) = 000...00, 00...01, 00...11....).

For example, consider the set of equations for a two-input system as follows:

$$\text{SET}_{Y_1} = AB + BY_2 \qquad \text{RESET}_{Y_1} = A'B' + B'Y_2 \qquad (9\text{-}31)$$
$$\text{SET}_{Y_2} = A'BY_1 + AB'Y_1 \qquad \text{RESET}_{Y_2} = ABY_1 + A'B'Y_1' \qquad (9\text{-}32)$$

Table 9-5 Computer Input Data Set

Card	0000000001111111111222222222233333333334444444445 1234567890123456789012345678901234567890
1	EXAMPLE PROBLEM
2	2 2 2
3	0000
4	1
5	2
6	1
7	2
8	MS(1) = ITS(3) + X(2)∗Y(2)
9	MR(1) = ITS(1) + NOT(X(2))∗Y(2)
10	MS(2) = ITS(2)∗Y(1) + ITS(4)∗Y(1)
11	MR(2) = ITS(3)∗Y(1) + ITS(1)∗NOT(Y(1))
12	Z(1) = ITS(2)∗NOT(Y(1))∗Y(2)
13	Z(2) = ITS(4)∗Y(1)∗Y(2)

NETWORK ANALYSIS AND REVISION

Table 9-6 Computer Output

Logic simulation
for 2 inputs, 2 outputs, 2 memories

Simulated flow table for
example problem

$X_1 X_2$ 00	10	11	01	Z_1	Z_2	Y_1	Y_2
(1)	2	0	3	0	0	0	0
1	(2)	4	0	0	0	0	0
1	0	4	(3)	0	0	0	0
0	6	(4)	5	0	0	1	0
1	0	4	(5)	0	1	1	1
1	(6)	4	0	1	0	0	1

$$Z_1 = A'BY_1'Y_2 \qquad (9\text{-}33)$$
$$Z_2 = AB'Y_1Y_2 \qquad (9\text{-}34)$$

Inserting $X(1)$ as A, $X(2)$ as B, and using $MS(1)$ as the SET signal of Y_1 while $MR(1)$ is its RESET signal, the program input data set can be constructed as shown in Table 9-5. Note the importance of recognizing AB as ITS(3), $A'B'$ as ITS(1), etc., which otherwise would be written as $X(1)*X(2)$ and $\text{NOT}(X(1))*\text{NOT}(X(2))$, respectively. The computer returns the flow table as shown in Table 9-6.

PROBLEMS

1 Derive the network equations for the implemented circuit in Figure P9-1 and compare the results with the desired network operation as represented disjunctively by the truth table shown.

Truth Table

a	b	c	d	e	f	z
0	0	0	0	0	0	0
0	–	0	1	1	0	1
0	0	0	1	1	–	1
–	0	–	–	–	1	1
0	1	0	1	1	1	0
–	–	1	–	–	1	1
1	1	0	0	0	0	1
1	1	0	0	0	1	0
1	1	0	0	1	1	1
–	–	–	0	–	1	1
–	–	–	–	0	1	1
1	1	0	0	1	0	0
1	–	–	–	–	1	1
1	1	1	–	–	0	0
0	–	1	–	–	0	0

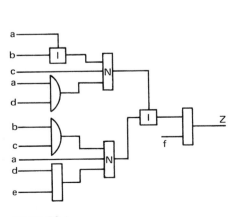

Figure P9-1

2 Compare the operation of the circuit in Figure P9-2 with the desired operation given by the Karnaugh map shown (assume disjunctive construction).

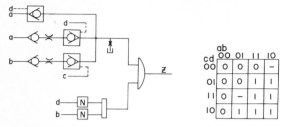

Figure P9-2

3 Using the most appropriate approach, correct the networks given in Problems 1 and 2.
4 Given the sequential network shown in Figure P9-4, construct the RSC of the representation. The network is designed to perform the logic required by a two-cylinder network having the input-output circuit shown in Figure 9-27 and is to perform the sequence: $A, \bar{A}, B, \bar{B}, AB, \overline{AB}$. . . . Determine the validity of the actual operation. If different, determine the actual operation of the machine.

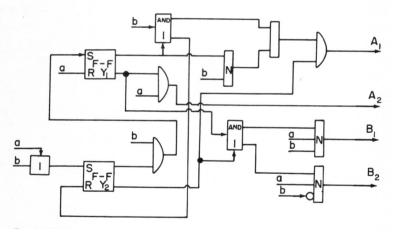

Figure P9-4

5 A machine is represented by the circuit equations given below:

$$\text{SET}_{Y_1} = ABY_2 + A'BY_3 \qquad \text{RESET}_{Y_1} = A'Y_2 + ABY_3 + A'B'$$
$$\text{SET}_{Y_2} = AB'Y_3' \qquad \text{RESET}_{Y_2} = A'Y_1' + Y_3$$
$$\text{SET}_{Y_3} = A'BY_2' + AB'Y_1 \qquad \text{RESET}_{Y_3} = B'Y_1'$$
$$Z_1 = ABY_1Y_3 + A'Y_2$$
$$Z_2 = A'B'Y_3 + B'Y_1'$$

Form the RSC of the network.

NETWORK ANALYSIS AND REVISION

6 By the classical revision methods, establish the corrections necessary for the networks given in Problems **4** and **5**.
7 Use the tabular means of analyzing the network given in Problem **4** and determine the necessary corrections.
8 The intuitively designed machine shown in Figure P9-8 is to be analyzed. The desired operation of the two-cylinder machines whose input-output circuits are given in Figure 9-27 is: $A, \bar{A}, A, \bar{A}, B, \bar{B}, B, \bar{B}$.... Utilize the most appropriate method for verifying the actual operation of the machine and make corrections where necessary.

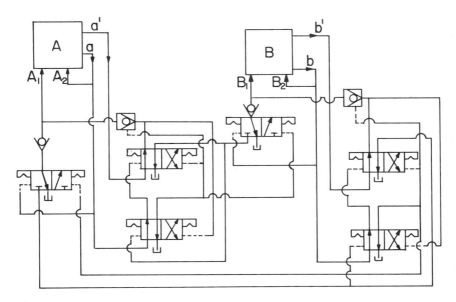

Figure P9-8

9 Verify the actual operation of the network shown in Figure P9-9. The input-output circuit of the network is given in Figure 9-27, while the intended machine cycle is: $A, B, \bar{B}, \bar{A}, B, A, \bar{A}, \bar{B}$.

DEFINITIONS

Analysis The examination of the individual elements of a control system and the determination of whether their integrated function satisfies a prescribed logic specification.

Auxiliary cycle A secondary cycle in a machine (or control system) which may occur during the operation of a machine (system).

Complete cycle The composition of all possible operations of a machine (system) including the principal cycle and all auxiliary cycles.

Feedback A condition where a secondary or primary output of a system becomes an input to the system itself.

Principal cycle The characteristic cycle of a machine (or system).

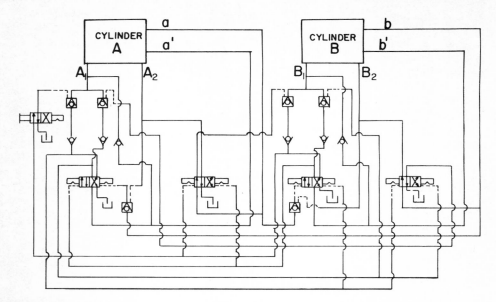

Figure P9-9

REFERENCES

Dietmeyer, D. L., *Logic Design of Digital Systems.* Boston: Allyn and Bacon, Inc., 1970.

Doig, G., and Walle, L. I., "Practical Air Circuitry—How to Design All-Air Control Systems for Automatic Machines," Manual Pac-65, Numatics Inc., Michigan, 1966.

Fitch, E. C., "Fluid Logic," Teaching Manual Published by The School of Mechanical and Aerospace Engineering, Oklahoma State University, Stillwater, Oklahoma, 1966.

_____, and Maroney, G. E., "The Operations Table Approach for Analyzing Digital Fluid Networks," Controls and Systems Conference, Chicago, Illinois, May 13-14, 1970.

_____, "Analysis of Fluid Circuit Networks," Fluid Logic Conference, Boston, Mass., April 16, 1970.

Givone, D. D., *Introduction to Switching Circuit Theory.* New York: McGraw-Hill Book Co., 1970.

Humphrey, W. S., *Switching Circuits With Computer Applications.* New York: McGraw-Hill Book Co., 1958.

Wood, P. E., *Switching Theory.* New York: McGraw-Hill Book Co., 1968.

Woods, R. L., "The State Matrix Method for the Synthesis of Digital Logic Systems," M. S. Thesis, Oklahoma State University, Stillwater, Oklahoma, 1970.

Anonymous, "Troubleshooting the Control Circuit Simulator," ARO Pneumatic Logic Controls Manual, Module 10, The ARO Corp., Ohio, 1973.

Appendix

A Typical Industrial Synthesis Problem

The main purpose of this appendix is to provide the reader with a clear overview of synthesizing a logic problem. The method utilized and demonstrated in this appendix is a tabular technique discussed in Chapter 8 and is called the Total Signal Method. Because of its simplicity and ability to resolve virtually all types of asynchronous sequential logic problems, it is highly recommended by the authors. To properly demonstrate the synthesis technique, an example problem introduced in Section 3-7 will be used (see Figure A-1). The problem description is reiterated again here for the convenience of the reader:

1. Reset the system by retracting cylinders A, B, and C, which is achieved by activating Z_2 and Z_6, and deactivating Z_3. Therefore, at this stage, inputs a, b, c, and d are activated; signal e is OFF and signals f and g are unknown or indeterminate.
2. If sensor g is activated, which indicates that a steel block is present, activate Z_1 until signal b is OFF.
3. After a delay of 0.5 seconds, activate cylinder C by the excitation of signal Z_5, and following another 0.5 seconds, start the drilling operation by the activation of signal Z_3.
4. Signal e indicates that the drilling operation is almost completed (that the depth of the hole has almost reached three inches). A mechanical limiting device is

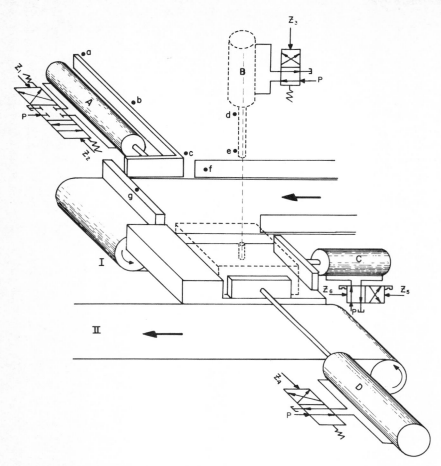

Figure A-1 An automatic drilling operation.

utilized to prevent the depth of the drilling from being over the intended measure; hence, a sufficient delay (e.g., one second) should be applied to signal e, which is then followed by the retraction of the drill by means of the deactivation of signal Z_3.

5 As soon as signal d is sensed, activate outputs Z_1, Z_4, and Z_6 simultaneously for transporting the block to the second conveyor. Signals Z_1 and Z_4 are held until signal c is OFF, to indicate that the block has fully reached the second conveyor.

6 Still holding the output signal Z_4 in the ON position (to permit the steel block to move downstream on the conveyor), retract cylinder A by activating Z_2 until signal a is sensed. Repeat the cycle by starting at Step 2.

7 If cylinder A has been extended, and signal f is activated during a later state, conveyor I is stopped to avoid any rubbing between the conveyor and the upstream steel blocks.

A TYPICAL INDUSTRIAL SYNTHESIS PROBLEM

As stated in Section 5-7, the synthesis table can be constructed using the following steps:

1. Arbitrarily select an initial condition which is used as state 1 of the Logic Specification. The starting operation of the system is accomplished by resetting the system. It is recommended that this starting state be selected as the first state of the logic system.
2. It is assumed that initially input signals f and g are not actuated, and a pressure switch is activated by output Z_7 to start the conveyor belt. All other outputs should be deactivated at this moment. Note, however, that the output element of cylinder C is a detented element; hence, the activation of output Z_6 would not interfere with the logic specification. A "-" ("don't care") output is assigned for this state to facilitate the simplification process. The outputs and inputs are listed in columns II and III of the synthesis table (Table A-1). For the convenience of the reader, a description of each state and the corresponding desired outputs are also given in this table.
3. The next step is to determine the next possible states of the machine as the cycle is triggered by the presence of a steel block; the first probable next state is the condition where a block passes sensor f. Since no action is to occur at this point, no changes in the outputs are made. This state is recorded as state 2 in Table A-1. The output state remains as in state 1 (e.g., $Z_1 Z_2 \ldots Z_7 =$ 0000-01), while the input state changes to $abcdefg = 1111010$. In column IV, row 1, a "2" is entered to indicate that state 2 is a possible next state of state 1.
4. Since there exists no other possible next states of state 1, the synthesis proceeds with the consideration of the next states of state 2. Initially, let it be assumed that the steel blocks are quite far apart, which implies the deactivation of input signal f. Again, no change of the outputs is expected.
5. When the steel blocks are positioned closely together, there exists a possibility that the input signal does not deactivate after the first block has passed location f. Under such circumstances, the next state is the case where signal g is activated (see state 4).
6. Similarly, there are two next states of state 3. In this case, however, it should be noted that one of its next states is one of its previous states. Note that recognizing existing states as possible next states is desirable in order to avoid unnecessary proliferation of the problem description.
7. The remaining states of the synthesis table are determined in a similar manner. The reader may realize the importance of detecting every possible next-state condition in order to insure a foolproof operation of the system.
8. After columns I-IV of the synthesis table have been completed, the previous possible states of each state are tabulated in the fifth column, as shown in Table A-2. Previous possible states are derived directly from the next-state information from column IV. For example, states 1, 3, and 24 indicate state 2 as a possible next state; hence, these states are the possible previous states of state 2. Similarly, states 6 and 9 are the previous possible states of state 8.
9. The next step is the augmentation of memory states. This augmentation is needed to establish the uniqueness of each state of the machine. This is performed by inspecting column III of the Synthesis Table of Table A-2. States

Table A-1 The Construction of the Synthesis Table

State	Outputs $Z_1 \ldots Z_7$	Inputs abcdefg	Next states	Physical condition	Expected action*
1	0-000-1	1111000	2	Start cycle, no block present	Activate conveyor l
2	0-000-1	1111010	3, 4	Block passing location f	Activate conveyor l
3	0-000-1	1111000	2, 5	Block has passed location f	Activate conveyor l
4	10000-0	1111011	6	One block aligned properly at g, another block present at f	Extend cylinder A, stop conveyor l
5	10000-1	1111001	4, 7	Block aligned properly at g	Extend cylinder A
6	10000-0	0111011	8	Input a deactivated due to extension of A, block present at f	Extend cylinder A, stop conveyor l
7	10000-1	0111001	6, 9	Input a deactivated	Extend cylinder A
8	10000-0	0111010	10	Block passes location g, another block present at f	Extend cylinder A, stop conveyor l
9	10000-1	0111000	8, 11	Block passes location g	Extend cylinder A
10	0010100	0011010	12	Block positioned on the drill bench, other block present at f	Extend cylinder C, activate Z_3 (delayed 0.5 and 1.0 sec), stop conveyor l
11	0010101	0011000	10, 13	Block positioned on the drill bench	Extend Cylinder C (delayed 0.5 sec), activate Z_3 (delayed 1.0 sec)
12	0010-00	0010010	14	Drilling under process, other block present at f	Continue activation of Z_3, stop conveyor l
13	0010-01	0010000	12, 15	Drilling under process	Continue activation of Z_3
14	0000-00	0010110	16	Desired drilling depth reached; other block present at f	Deactivate Z_3, (delayed 1.0 sec), stop conveyor l
15	0000-01	0010100	14, 17	Desired drilling depth reached	Deactivate Z_3 (delayed 1.0 sec)
16	0000-00	0010010	18	Drill bit retracting, another block present at f	Deactivate Z_3, stop conveyor l
17	0000-01	0010000	16, 19	Drill bit retracting	Deactivate Z_3
18	1001010	0011010	20	Drill bit fully retracted, another block present at f	Extend cylinder A, retract cylinders C and D, stop conveyor l
19	1001011	0011000	18, 21	Drill bit fully retracted	Extend cylinder A, retract cylinders C and D
20	01010-0	0001010	22	Cylinder A fully extended, another block present at f	Retract cylinders A and D, stop conveyor l
21	01010-1	0001000	20, 23	Cylinder A fully extended	Retract cylinders A and D
22	01010-0	0011010	24	Cylinder A retracting, another block present at f	Retract cylinders A and D; stop conveyor l
23	01010-1	0011000	22, 25	Cylinder A retracting	Retract cylinders A and D
24	01010-0	0111010	2	Cylinder A retracting, another block present at f	Retract cylinders A and D; stop conveyor l
25	01010-1	0111000	1, 24	Cylinder A retracting	Retract cylinders A and D

Table A-2 The Completed Synthesis Table

State	Outputs $Z_1 \ldots Z_7$	Inputs $abcdefg$	Next states	Prev. states	Memory augmentation	Y_1	Y_1^*	Y_2	Y_2^*	Y_3	Y_3^*	Y_4	Y_4^*	Y_5	Y_5^*	Y_6	Y_6^*	Y_7	Y_7^*	Y_8	Y_8^*	Y_9	Y_9^*
1	0–000–1	1111000	2	25	y_1'	0	0																
2	0–000–1	1111010	3, 4	1, 3, 24		1	S																
3	0–000–1	1111000	5, 2	2	y_1	1	1																
4	10000–0	1111011	6	2, 5																			
5	10000–1	1111001	4, 7	3																			
6	10000–0	0111011	8	4, 7				0	R														
7	10000–1	0111001	6, 9	5						0	R												
8	10000–0	0111010	10	6, 9	y_2'			0	0														
9	10000–1	0111000	8, 11	7	y_3'			0	R	0	0												
10	0010100	0011010	12	8, 11	$y_4'y_5'$							0	0	0	0								
11	0010101	0011000	10, 13	9	$y_6'y_7'$							0	R	0	R	0	0	0	0				
12	0010–00	0010010	14	10, 13	y_8'															0	0		
13	0010–01	0010000	12, 15	11	y_9'															0	R	0	0
14	0000–00	0010110	16	12, 15																1	S		
15	0000–01	0010100	14, 17	13																		1	S
16	0000–00	0010010	18	14, 17	y_8							1	S			1	S						
17	0000–01	0010000	16, 19	15	y_9									0	R	1	1	0	R				
18	1001010	0011010	20	16, 19	y_4y_5							1	1	0	0	1	1	0	0				
19	1001011	0011000	18, 21	17	y_6y_7'							1	S	0	0	1	1	1	S				
20	01010–0	0001010	22	18, 21														1	1				
21	01010–1	0001000	20, 23	19				1						1	1								
22	01010–0	0011010	24	20, 23	y_5									1	S								
23	01010–1	0011000	22, 25	21	y_7			1	1									1	1				
24	01010–0	0111010	2	22, 25	y_2			1	S														
25	01010–1	0111000	1, 24	23	y_3	0	R	1	1	1	S												

Memory SET/RESET

having the same input state are grouped and augmented by a sufficient number of memory elements—for N states, at least $\log_2 N$ memory devices are needed. Using these memory devices, N combinations of memory signals (memory states) are formed, preferably in Gray Code fashion. For example, input states 1 and 3 are equal; therefore, as there are only two states involved, only one memory variable is required. By arbitrarily assigning y'_1 to state 1 leaves no alternative than to assign y_1 to state 3. Similarly, states 8 and 24 have identical input states, and hence, y'_2 and y_2 are assigned to these states to establish their uniqueness.

10 A slightly different situation occurs when more than two states are involved. A four-state equivalency condition would require two memory variables; augmented with the following Gray Code form: $y'_m y'_n$, $y'_m y_n$, $y_m y_n$, $y_m y'_n$. For situations where more memory states are available than are needed, more than one memory state can be assigned to one or more of the machine states. The main objective of such multiple assignment is the simplification of the augmented memory states; therefore, grouping of the memory states should be performed such that they can be simplified. For example, states 10, 18, and 22 have equal input states; and as $\log_2(3)$ is 1.585, two memory elements are required. Having assigned $y'_4 y'_5$ and $y_4 y'_5$ to states 10 and 18, respectively, two memory states still remain, and as these memory states simplify to y_5, state 22 is assigned to y_5. Similarly, a six-state equivalency would require three memories, which is best assigned as follows: $y'_i y'_k y'_n$, $y'_i y'_k y_n$, $y_i y'_k y_n$, $y'_i y_k y'_n$, $y_i y'_k$, and $y_i y_k$.

11 After each machine state has been properly augmented, it is necessary that the required conditions of each memory element be determined and recorded properly. For this purpose, two columns in the table are allocated to each of the memory elements. In Table A-2, these columns are designated by columns Y_n and Y_n^*. In the Y_n column, the required binary-state condition of memory Y_n is recorded. For example, in the Y_1 column, place 0's for each state where y'_1 is needed as well as all the preceding states. In Table A-2, for example, these states are states 1 and 25. Similarly, 1's are inserted in the table for states where y_1 is required and all the preceding states (see states 2 and 3).

12 After all Y columns have been completed, the SET/RESET states of each memory element are determined and recorded in the Y_n^* columns. For the RESET state, locate the 0— states of the Y_n column which have an entry other than 0 in one or more of their previous states. These 0 — states are assigned R's to denote RESET states. Similarly, all 1-states, which have entries other than 1's, in one or more of their previous states, are designated as S's to signify SET states. For example, state 2 is the SET state of memory Y_1, because its previous states (as observed in the fifth column), (e.g., states 1 and 24) do not have 1-states. Similarly, state 25 is the RESET state, since its previous state, state 23, does not have a "0" y_1 state.

13 Steps 11 and 12 are performed iteratively until all memory variables are considered.

14 At this stage, the sequential phase of the synthesis has been completed. Each combination of the input state and its associated memory state (if any) is a unique representation of the machine state. Therefore, each output (both primary or secondary) can be correctly represented by these input-memory combinations. For example, since output Z_1 is represented in only states 4

A TYPICAL INDUSTRIAL SYNTHESIS PROBLEM

through 9 and in states 18 and 19, output Z_1 can be described by the following expression:

$$Z_1 = abcde'fg + abcde'f'g + a'bcde'fg + a'bcde'f'g + a'bcde'fg'y'_2 \\ + a'bcde'f'g'y'_3 + a'b'cde'fg'y_4y'_5 + a'b'cde'f'g'y_6y'_7$$

Similarly, the expression for the SET signal of Y_1 is:

$$\text{SET}_{Y_1} = abcde'fg'$$

15. The final step is the simplification of the network equations. Various simplification methods have been presented in Chapter 6 for combinational networks. For problems having more than six variables, Karnaugh maps should not be applied. The use of computer-aided simplification is a necessity in such cases. As the particular problem considered in this Appendix has sixteen variables (inputs and memories), manual simplification is out of the question. The results of the computer simplification is presented below:

$$Z_1 = g + a'bfy'_2 + a'bf'y'_3 + b'cdfy_4y'_5 + b'cdf'y_6y'_7$$
$$Z_2 = c' + b'dfy_5 + b'df'y_7 + bfg'y_2 + bf'g'y_3$$
$$Z_3 = b'cdfy_4 + b'df'y_6 + b'd'e'fy_8 + b'd'e'f'y_9$$
$$Z_4 = b'df(y_4 + y_5) + b'df'(y_6 + y_7) + c' + a'bfg'y_2 + a'bf'g'y_3$$
$$Z_5 = b'cfy'_4y'_5 + b'cf'y'_6y'_7$$
$$Z_6 = dfy_4 + df'y_6$$
$$Z_7 = f' + ag'$$

The memory SET equations are simplified by assigning the S states as 1's while all 0's and R's are used as the 0's of the SET expression. Similarly, the simplified memory RESET equations are derived by using the R states as 1's while all 1's and S's constitute the 0's of the RESET expression. All other states are used as the "don't cares" of the particular expression. Hence, the simplified SET and RESET equations are as follows:

$$S_{Y_1} = f \qquad\qquad R_{Y_1} = a'$$
$$S_{Y_2} = b' \qquad\qquad R_{Y_2} = b$$
$$S_{Y_3} = b' \qquad\qquad R_{Y_3} = g$$
$$S_{Y_4} = d' + f'y_6 \qquad\qquad R_{Y_4} = b + f'y'_6$$
$$S_{Y_5} = c' + f'y_7 \qquad\qquad R_{Y_5} = b + d' + f'y'_7$$
$$S_{Y_6} = d' \qquad\qquad R_{Y_6} = b$$
$$S_{Y_7} = c' \qquad\qquad R_{Y_7} = b + d'$$
$$S_{Y_8} = e + f'y_9 \qquad\qquad R_{Y_8} = d + f'y'_9$$
$$S_{Y_9} = e \qquad\qquad R_{Y_9} = d$$

It should be obvious to the reader that the synthesis can be divided into three distinct phases, as follows:

1. Formulating of the problem specification.
2. The sequential portion of the synthesis.
3. The combinational portion of the synthesis.

Oftentimes, the formulation of the problem is the most critical phase in the resolution of a logic problem, because at this stage, correct judgments regarding the evaluations of each possible condition are made. The integrity of the control network depends entirely upon the interpretation and degree of perfection applied by the designer. Since the optimality of the network is often achieved at the expense of simplicity, compromises are usually necessary. Unfortunately, the formulation of the problem specification can only be performed manually, as no computer or other mechanical means has been developed for this particular purpose.

The sequential portion of the synthesis is the process of transforming the sequential problem into a set of combinational problems. In particular, the process includes the construction of the synthesis table and the assignments of the memory variables. The combinational synthesis portion of the synthesis reformulates and simplifies the network equations using various techniques as discussed in Chapter 6 of this book. The major motivation for applying simplification methods to combinational networks is to enhance their practicality and economy, since simplified equations means easier implementation and less utilization of fluid logic hardware. Due to the fact that the sequential and combinational synthesis processes follow a set of precisely defined rules, this phase should not create any critical problems to the designer. Moreover, for larger logic problems, effective computer programs are available for performing these tasks.

Definitions

Accessory element A device that is not a logic element but is needed to implement a complete logic circuit.
Active element A logic element that has one of its passage ports connected to a power source.
Actuator A device for moving or controlling something indirectly.
Adjacency A relationship existing between two stable states such that there is only one describing variable which is different in value (0, 1).
Air logic Relating to fluid logic components operated by air.
Analysis The examination of the individual elements of a control system and the determination of whether their integrated function satisfies a prescribed logic specification.
Auxiliary cycle A secondary cycle in a machine (or control system) which may occur during the operation of a machine (system).
Circuit An assemblage of elements and their interconnection.
Class A group sharing common conditions.
Combinational circuit A logic circuit in which the output at any given instant is dependent upon the present inputs.
Complement The mathematical inverse of the entity. Two quantities or symbols are complements of each other if and only if one quantity or symbol takes the value of one when the other equals zero, and vice versa.

Complementary expression An expression whose literals, operators, and constants have inverted forms, i.e., 0, 1, +, ·, complemented literals and uncomplemented literals in one expression correlates with 1, 0, ·, +, uncomplemented literals, and complemented literals in its complementary expression.

Complete cycle The composition of all possible operations of a machine (system) including the principal cycle and all auxiliary cycles.

Conjunctive canonical form A conjunction of maxterms.

Conjunctive expression A Boolean expression that contains a single sum term or a product (conjunction) of sum terms.

Conjunctive form A product (conjunction) of sum terms.

Constant A symbol that represents the values of the algebra and can be only 0 or 1.

Control medium The medium used to transmit a process and amplify signals.

Control network A network designed to analyze and formulate the operations of a machine.

Control point of a valve The actuation port of a valve.

Control system A system consisting of the control network, input sensors, and outputs.

Deterministic circuit A logic circuit in which only predetermined inputs are allowed.

Digital Relating to discrete operations or systems.

Disjunctive canonical form A disjunction of minterms.

Disjunctive expression A Boolean expression that contains a single product term or a sum (disjunction) of product terms.

Disjunctive form A sum (disjunction) of product terms.

Don't care output An output which is not required to satisfy the system function.

Don't care terms Minterms and maxterms which can be added to a function without affecting its desired operation.

Dual expressions Expressions whose operators and constants have inverted forms, i.e., 0, 1, +, and · in one expression correlates with 1, 0, ·, and + in its dual, respectively.

Element A device used to achieve a given function.

Equation An equality involving two or more constants, variables, or functions.

Equivalent expressions Expressions which have the same value when the values of the variables are equal.

Event One discrete stage in a sequence.

Expression A combination of variables and operators.

Fan-in The number of permissible logical inputs to a logic element.

Fan-out The number of logic elements of like kind that can be operated or controlled by the given element when all elements are operating at the same pressure.

Feedback A condition where a secondary or primary output of a system becomes an input to the system itself.

Feedback inputs Inputs that indicate the completion of a certain machine operation.

Flow passage of a valve A path that internally connects one port to another, at one or more, but not all, actuating positions of the valve.

Fluidic Relating to nonmoving part fluid elements.

Fluid logic The study of a system utilizing digital control elements and circuits that uses fluids (liquids or gases) to transmit logic signals.

DEFINITIONS

Function A mathematical expression describing the relation between variables.

Hazard An actual or potential network malfunction as a result of network delays or imperfections.

Hydra-logic Relating to fluid logic components operated by liquids.

Imperfect logic element A logic device that satisfies a part of the logical relationships stated by the prescribed truth table.

Implementation The act or process of representing an algebraic description of a network by means of hardware application.

Implicant of a function A product term of a function that does not contain any pair of complementary literals (e.g., aa').

Implicate of a function A sum term of a function that does not contain any pair of complementary literals (e.g., $a + a'$).

Indeterminate variable value A variable value that can be any of the two values of the algebra.

Input A signal used to represent the state of a particular condition.

Input sensor A device used to produce inputs.

Interface A device that serves as a boundary between two different signal transmission media.

Internal state An explicit description of the conditions of the memories.

Level or stage location of an element The number of elements through which the input signals of the element under consideration must transgress in order to reach the output.

Levels or stages in a network The largest or maximal stage location in a network.

Literal An individual entry of a variable into an equation in either its complemented or its uncomplemented form.

Logic Relating to reasoning and decision-making.

Logic element A device that is capable of making a TRUE or FALSE, YES or NO, "1" or "0" output decision, based upon its input condition.

Logically complete element or operator An element or operator that can satisfy all logic functions either by itself or with the conjunction of equivalent elements or operators.

Machine cycle A course of operations of the machine that recur regularly.

Machine event A describable stage in the operating sequence of a machine.

Machine state An explicit description of the machine event.

Maxterm A sum term that contains every variable in the system in either its complemented or its uncomplemented form.

Memory excitation A stimulation of the secondary or memory system.

Memory or secondary state The collective state or condition of the memory elements.

Memory transition The stepwise change of the memory state.

Minimal Row Specification Chart A specification chart that describes the logic operation of a machine with the minimal number of rows.

Minterm A product term that contains every variable in the system in either its complemented or its uncomplemented form.

Multiple-output prime implicant or implicate An implicant or implicate that represents one or more outputs and that is irreducible for the representation of the particular combination of outputs.

N-way valve A valve exhibiting N ports for directing fluid.

Network An assemblage of circuits and their interconnection.

Nonlogic elements Elements which do not perform a logic function.
Normal form A disjunctive form or a conjunctive form.
Operation A defined action.
Operational Flow Chart A specification chart that represents the operational logic of a machine and which satisfies the adjacency requirements of its memory states.
Operator A symbol that denotes a mathematical operation.
Output A signal used to represent the external state of a system.
Passive logic element A logic element that has none of its ports connected to a power source.
Perfect logic element A logic element that satisfies all logical relationships stated by the prescribed truth table.
Predetermined input An input generated as a result of a fixed sequence of events.
Prepared path condition A condition where the change of the input causes a desired change of the output without any switching of logic elements.
Primary input An input from an external source to a control network source in a medium compatible with the network.
Primary output An output that is used for the control of a machine.
Prime implicant of a function An irreducible implicant of a function.
Prime implicate of a function An irreducible implicate of a function.
Principal cycle The characteristic cycle of a machine (or system).
Product term A conjunction of literals, a group of literals interconnected by AND's.
Random input An input occurring in a totally unpredictable or probabilistic manner.
Reduced Specification Chart A logic specification chart containing the minimal number of machine states.
Row compatibility Two rows are compatible if they can be merged into one row.
Secondary input An input produced by the control network itself.
Secondary output An output used to reflect the present state of a machine.
Sensor An element which transforms a state property into a signal.
Sequential Occurring as a result of present and past states.
Sequential circuit A logic circuit in which the output at any given instant is dependent not only upon the present inputs but also on the input history.
Signal A detectable physical quantity or impulse representing the state of a system.
Stable state A machine state that does not change without a change in the inputs.
State A description of the condition, stage, position, level, etc., of a physical system.
State equivalency Two machine states are equivalent if they can be represented by one machine state.
Stochastic circuit A logic circuit in which predetermined, random, or both types of inputs are permissible.
Sum term A disjunction of literals, a group of literals interconnected by OR's.
Synthesis The act or process of resolving a logic description and determining the best algebraic description.
System A collection of elements united by some form of interaction.
Unary element An element that has one input.
Unstable state A transitory condition of the machine during which a machine attempts to reach an intended machine state.
Variable The various letters and/or symbols that are not operators.
Variable value A constant that represents a numerical condition of a variable.

Index

Absorption, 12
Accessory element, 112 (def.)
   symbols for, 94
Accounting table, machine state, 193
Acoustic sensor, 47, 51
Active element, 76, 95, 112 (def.)
Actuator, 43, 67 (def.)
   ball, 46
   manual type, 43, 46
   push button, 43, 46
   roller lever, 43, 44
   roller plunger, 43, 45
   toggle, 43, 46
   wire rod, 46
Addition of rows, 200
Adjacency, 26, 231 (def.)
   connection, 204
   graph, 202
   on Karnaugh map, 203
   mandatory, 202

Adjacency (*Cont.*):
   primary, 201
   secondary, 201
Adjacent rows, 199, 201
   mandatory, 202
Air logic, 2, 8 (def.)
American National Standards Institute, 70
   fluid diagrams, 105
   fluid symbols, 70
   input symbols, 103
   logic symbol, 70
   output symbols, 104
   peripheral equipment symbols, 104, 105
Amplification, input, 48, 52, 53
Amplifier:
   closed, 88, 89
   element, 88
   open, 88, 90

INDEX

Analysis, 275, 307 (def.)
    combinational, 276
    computer program, 304
    logic, 275
    network, 275
    sequential, 284
    table, 298
AND:
    as basic operation, 11, 14
    element, 71
    INHIBITOR element, 83
    NOR construction, 145
    NOT construction, 144
    OR element, 72, 73
A.N.S.I. (*see* American National Standards Institute)
Applications:
    agricultural, 6
    aircraft, 5
    biomedical, 5
    industrial, 5
    machine tool, 6
    packaging, 5
Assignment:
    memory state, 200, 208
    secondary state, 200, 208
Association, 12
Augmentation, W shut-off valves, 238
Auxiliary cycle, 307 (def.)
Auxiliary memory, 236

Back-pressure sensor, 46, 49
Ball actuator, 43, 46
Bar chart, 116
Bensch, L. B., 267
Binary code, 24
Binary designation of states, 11, 208
Blocked center valve, 43
Boole, G., 2
Boolean algebra, 2, 10
    postulates, 11
    theorems, 11
Burchett, O. J., 3

Cam-follower system, 56, 57
Canonical flow table, 270
Canonical form, 20, 22
Cell, 25

Change Signal Method, 235
Change signal program, 267
Chart, excitation, 207
Check valve logic, 278
Chen, P. I., 259
Cheng, R. M. H., 253
Circuit, 8 (def.)
    minimal memory, 200
Class, 13, 38 (def.)
Classical synthesis, 186
CLASYN program, 226
Closed amplifier, 88, 89
Closure, 194
Coincidence element, 81
Cole, J. H., 3, 97, 119, 235
Combinational analysis, 276
Combinational circuit, 2, 8 (def.)
Combinational logic system, 137
Combinational network hazard, 166
Common factor, 150
Commutation, 12
Compatibility graph, 196
Compatibility of rows, 196, 231 (def.)
Compatible machine states, 196
Complement, 38 (def.)
Complementary entry, 30
Complementary expression, 11, 38 (def.), 282
Complementary function, 11, 38 (def.), 282
Complementary reset point, 240
Complementation, 11, 12
    program, 176
    reduction by, 143
Complete cycle, 307 (def.)
Complete polygon, 191
Completeness:
    functional, 155
    logical, 153
Complications, logical, 217
Conflicting condition, 235
Conflicting events, 238
Conflicting input-output, 235
Conjunctive canonical form, 22, 38 (def.)
Conjunctive expression, 21, 38 (def.)
Conjunctive form, 21, 38 (def.)
Constant, 38 (def.), 39
Contractive element, 155

INDEX

Contraposition, 12
Control:
    flow passage symbols, 93
    medium, 67 (def.)
    network, 40, 67 (def.)
    point, 91
        symbols, 91
        of a valve, 112 (def.)
    system, 40, 67
Conventional solution, 138
Correction, 276
Critical:
    PI, 169
    race, 218
    row, 203
    transition, 199, 207
Cut set, 177
Cycles, 221
    complete, 307 (def.)
    principal, 307 (def.)

d-trio, 224
DeMorgan's theorem, 12, 144
Design, 137
Detent power system, 56
Detented memory, 80, 214
Deterministic circuit, 115, 135 (def.)
Diagram:
    attached, 108
    detached, 91, 106
Diagramming:
    attached, 101
    detached, 102
    in synthesis, 259
Diconesyn III, 126
Digital, 1, 8 (def.)
Disjunctive canonical form, 22, 38 (def.)
Disjunctive expression, 21, 38 (def.)
Disjunctive form, 21, 38 (def.)
Distribution, 12
"Don't care," 140
    terms, 30, 38 (def.), 140
    output, 133, 135 (def.)
Double negation, 12
Dual expressions, 11, 38 (def.)
Dynamic function hazard, 170
Dynamic hazard, 167

Element, 8 (def.)
    active, 76, 112 (def.)
    unary, 74, 112 (def.)
EPC, 188
Equation, 11, 38 (def.)
    implementation, 97
Equivalency:
    graph, 191
    of states, 188, 231 (def.)
Equivalent Pairs Chart, 188
Equivalent set, 191
    machine states, 191
Equivalent states, 188
Essential hazards, 222
Essential sets, 193, 196
Event, 135 (def.)
    conflicting, 238
Excitation:
    chart, 207
    maps, 208
        Karnaugh, 210
        preliminary, 289
    memory, 231 (def.)
    modification of, 294
Exclusive OR:
    coincidence element, 84
    element, 81
Expanded output subtable, 122
Expansion, 12
Expression, 11, 38 (def.)
    equivalent, 11, 38 (def.)
External inputs, 255

Factor, common, 150
Factoring, 145
False output, 217
    hazard, 221
Fan-in, 96, 112 (def.)
    limits, 179
    multi, 160
    unlimited, 160
Fan-out, 96, 112 (def.)
    limits, 179
Favorable SET/RESET, 236
Feedback, 307 (def.)
    inputs, 255, 273 (def.)
    type input, 255
    type networks, 235

Fitch, E. C., 3
Flip-flop:
    fluidic, 79
    memory, 79
Flow chart, operational, 199, 231 (def.)
Flow passage, 112 (def.)
Fluid logic, 1, 8 (def.)
FLUID MEMORY, 77
Fluid sensors:
    four-way, 43
    three-way, 43
Fluidic, 2, 8 (def.)
Fluidic Flip-Flop, 79
Foster, K., 253
Four-way fluid sensor, 43
Function, 38 (def.)
    hazard, 169
Functionally complete element, 155
Fundamental mode, 186

Graph, adjacency, 202
Gray code, 24, 236

Hazard, 166, 183 (def.)
    eliminating term, 174
    essential, 222
    logic system, 166
    Nth order, 225
    second order, 225
    sequential, 217
    static type, 167
    steady state, 217
    third order, 225
    transient, 225
Homing in procedure, 303
Huffman, D. A., 3, 120, 186
Hydra-logic, 8 (def.)

Imperfect logic element, 112 (def.)
Implementation, 97, 183 (def.)
    equation, 97
    multiple output, 151
    special memory, 211
Implicant of a function, 38 (def.)
Implicate of a function, 38 (def.)
Indeterminate logic operation, 199

Indeterminate variable value, 34, 38 (def.)
Inhibiting function, 281
INHIBITOR:
    element, 82
    type function, 159
Input, 8 (def.)
    external, 255
    feedback type, 255
    interface, 48
    memory race, 223
    sensor, 40, 41, 67 (def.)
Interface, 40, 48, 67 (def.)
    electric to fluid, 47, 52
    fluid to electric, 48, 52
    fluid to fluid, 48, 52, 53
    hydraulic to pneumatic, 48
    input, 48
Internal state, 67 (def.)
Interruptible jet sensor, 47, 50
Irregularly activated network, 245

Karnaugh, M., 24, 120
Karnaugh map, 24
    adjacency, 203
    excitation, 210

Ladder diagrams, 91, 106
Lee, Y. H., 259
Level, 155, 183 (def.)
    of an element, 183 (def.)
    in a network, 183 (def.)
Literal, 12, 38 (def.)
Load-detection system, 56
Lockup, 221
Logic, 8 (def.)
    analysis, 275
    element, 70, 112 (def.)
        active, 76, 112 (def.)
        imperfect, 112 (def.)
        passive, 76, 95, 112 (def.)
        perfect, 71, 112 (def.)
        universal, 91
Logic specification, 116
Logic Specification Chart, 127, 187
Logic system description, 114
Logical complications, 217

INDEX

Logically complete element, 153, 183 (def.)
Logically complete operator, 183 (def.)
LSC, 127, 187

Machine:
   cycle, 67 (def.)
   event, 135 (def.)
   state, 67 (def.), 120, 135 (def.)
      accounting table, 193 .
Mandatory adjacency, 202
Mandatory memory, 235
Manual-type actuators, 43, 46
Maps:
   excitation, 207
   output, 215
Maroney, G. E., 3, 126, 245
Matrix:
   format, 164
   method, 250
   state, 124
Maximal equivalent set, 191
Maxterm, 22, 38 (def.)
McCluskey, E. J., Jr., 33
Memory:
   auxiliary, 236
   detented, 80, 214
   element, 77
   excitation, 231 (def.)
   flip-flop, 79
   fluid, 77
   mandatory, 236
   powerless, 278
   state, 200, 231 (def.)
      assignment, 200
   transition, 231 (def.), 255
Mergeable rows, 196
Merged specification chart, 197, 257
Minimal memory circuit, 200
Minimal operation flow chart row, 200
Minimal Row Specification Chart, 195, 197, 231 (def.)
Minimal truth table, 138
Minterm, 22, 38 (def.)
MOMIN computer program, 153, 178
Moore, E. A., 3
MSC, 195, 197, 231 (def.)

Multi-fan-in elements, 160
Multiple output:
   circuit, 150
   network for implementation, 150
   prime implicant, 183 (def.)
   prime implicate, 183 (def.)
   simplification, 249
Multiterminal networks, 148

N-level network, 156
N-way valve, 43, 67 (def.)
Nth-order hazard, 225
NAND:
   element, 74
   synthesis program, 179
National Fluid Power Association fluidic symbol, 70
Near minimal Operational Flow Chart, 201
Near minimal state machine, 191
Network, 8 (def.)
   analysis, 275
   deterministic, 235
   feedback type, 235
   irregularly activated, 245
   stochastic, 245
Next-state coverage, 194
Next-state subtable, 122
Next-state table, 193
Nodal diagram, 191
Node:
   principal, 202
   secondary, 202
Nonclassical synthesis, 234
Noncritical race, 218
   insertion of, 219
Nonequivalent states, 189
Nonlogic elements, 130, 131, 135 (def.)
NOR:
   computer program, 142, 176
   element, 76, 155
   operator, 156
   passive type, 160
   three level, 159
Normal form, 21, 38 (def.)
NOT:
   as basic operation, 11, 14
   element, 74
Null class, 12

OFC (*see* Operational Flow Chart)
"Off" terms, 34
One-shot element, 85
Open amplifier, 88, 90
Open center valve, 43
Operation, 8 (def.)
Operational Flow Chart, 199, 231 (def.)
   minimal row, 200
   near minimal, 201
Operations table, 119, 235
   method, 119
Operator, 11, 38 (def.)
Optional entry, 30
OR:
   as basic operation, 11, 14
   element, 72
Output, 8 (def.)
   circuits, 54
   equations, 215
   maps, 215
      development of, 287
   persistent, 236
   subtable, 122

Passage, controlled flow, 91
Passive logic element, 76, 95, 112 (def.)
Passive NOR element, 160
Perfect logic element, 71, 112 (def.)
Persistent outputs, 236
PFT, 120, 186
Pilot check valve system, 56, 58
Poppet valve-type sensor, 45, 48
Position-sensing system, 56, 57
Power circuits, 54
Powerless memory, 278
Predetermined input, 115, 135 (def.)
Preliminary excitation map, 289
Preliminary state matrix, 251
Prepared path, 97, 235
Prepared signal path, 235
Pressure signal valve, 44, 47
Primary adjacency, 201
Primary input, 41, 67 (def.)
Primary output, 41, 67 (def.)
Prime implicant 26, 38 (def.)
Prime implicate, 26, 38 (def.)
"Prime" term, 26, 38 (def.)
Primitive Flow Table, 120, 186

Principal cycle, 307
Principal node, 202
Product term, 20, 39 (def.)
Proximity sensor, 46, 50
Push button actuator, 43, 46

Quine, W. V., 33

Race, 199, 218
   critical, 218
   input-memory, 223
   noncritical, 218
Random input, 115, 135 (def.)
R-C network, 172
Reduced Specification Chart, 194, 231 (def.)
Redundant machine state, 188
Reference form, 138
Reflection, 12
Relay:
   actuating, 89
   porting, 89
   symbology, 89, 91-95
   valves, 89, 90
RESET:
   equations, 212
   favorable, 236
   unavoidable, 236
Response time, 4, 155
Resynthesis, 276
Revision, 279
   combinational, 279
   sequential, 293
Revolution sensor, 46, 49
Roller-lever actuator, 43, 44
Roller-plunger actuator, 43, 45
Ronan, H. R., 3
Rows:
   addition approach, 200
   adjacent, 199-201
   arranging procedure, 204
   compatibility, 196, 231 (def.)
   critical, 203
   mandatory, 202
   minimization, 196
RSC, 194, 231 (def.)

Second-order hazard, 225
Secondary adjacency, 201
Secondary map, development of, 287
Secondary input, 41, 67 (def.)
Secondary node, 202
Secondary output, 41, 67 (def.)
Secondary state, 231 (def.), 298
   assignment, 208
Sensor, 8 (def.)
   acoustic, 47, 51
   back pressure, 46, 49
   four-way valve, 43
   interruptible jet, 47, 50
   poppet valve, 45, 48
   proximity, 46, 50
   three-way valve, 43
   wire, 43, 46
Sequence programmers, 51, 54
Sequential, 3, 8 (def.)
Sequential analysis, 284
Sequential circuits, 3, 8 (def.)
Sequential network:
   hazards, 217
   revision, 293
SET:
   equations, 212
   favorable, 236
   unavoidable, 236
Shannon, C. E., 2
Sharping, 281
Shutoff valves, 236
Signal, 8 (def.)
   scheduling, 223
Simplification:
   multiple output, 148, 249
   network, 137
   program, 172
Smoothing, 172
Special memory implementation, 211
Specification chart:
   merged, 197, 257
   minimal row, 195, 197, 231 (def.)
   reduced, 194, 231 (def.)
Speed-governing system, 56, 58
Speed of response, 155
Stability characteristics, 121
Stable:
   condition, 186, 208
   state, 120, 136 (def.)

Stage location of an element, 183 (def.)
Stages, 96, 155
   in a network, 183 (def.)
State, 8 (def.), 112 (def.)
   binary designations, 208
   equivalency, 188, 231 (def.)
   input, 186
   memory, 186, 231 (def.)
   secondary, 208, 231 (def.)
   substitution, 191
   unstable, 120, 136 (def.), 208
State Diagram Method, 259
State Matrix Method, 124, 250
   preliminary, 251
Static:
   hazard free, 174
   1-function hazard, 170
   1-hazard, 168
   1-logic hazard, 169
   0-function hazard, 170
   0-hazard, 169
   0-logic hazard, 169
Steady-state hazard, 217
Stochastic circuit, 115, 136 (def.)
Stochastic synthesis method, 245
Subcube, 25
Subtable:
   next state, 122
   output, 122
Sum term, 20, 39 (def.)
Supplemental set, 193
Surjaatmadja, J. B., 3, 33
Synthesis, 183 (def.)
   classical, 186
   nonclassical, 234
   table, 126, 235, 246
System, 8 (def.)

TAB II, 33, 172
Tabular analysis, 298
Tautology, 12
Term, 20
   product, 20, 39 (def.)
   sum, 20, 39 (def.)
Third-order hazard, 225
Three-level network, 156
Three-level NOR logic, 159
Three-way fluid sensor, 43

Thrust vector, logic design, 7
Tie-set, 177
Time delay:
    circuit, 53
    element, 87
Timing chart, 116
Timing devices, 51, 53
TIMING-IN element, 87
TIMING-OUT element, 87
Toggle actuator, 43, 46
Total Signal Method, 245
Total signal program, 267
Transient hazard, 225
Transition, 12, 199
    memory, 255
Transition Table Method, 253
Transitions:
    critical, 199, 207
    memory, 231
Transposition, 12
Troubleshooting, 302
Truth table, 12, 115
    minimal, 138
    proof, 12
Two-valued algebra, 11

Unary element, 74, 112 (def.)
Unavoidable SET/RESET, 236
Uncomplementation, 23
Unger, S. H., 222
Universe class, 12
Unlimited fan-in, 160
Unstable state, 120, 136 (def.), 208

Valve:
    shutoff, 236
    solenoid operated, 48, 52
Variable, 12, 39 (def.)
Variable value, 11, 39 (def.)
Venn diagram, 13

W-valve, 236
    augmentation, 238
Wire-rod actuator, 46
Wire sensor, 43, 46
Woods, R. L., 3, 124, 250

"Yes" valve, 172